Atlas of
Industrializing Britain
1780–1914

Atlas of
Industrializing Britain
1780–1914

Edited by **John Langton** *and* **R.J. Morris**

Methuen   London and New York

First published in 1986 by
Methuen & Co. Ltd
11 New Fetter Lane, London EC4P 4EE

Published in the USA by
Methuen & Co.
in association with Methuen, Inc.
29 West 35th Street, New York, NY 10001

Typeset in Monophoto Century by
Vision Typesetting, Manchester

Printed in Great Britain at the
University Press, Cambridge

British Library Cataloguing in Publication Data

Atlas of industrializing Britain 1780–1914.

1. Great Britain – Industries – History – Maps
I. Langton, John.   II. Morris, R.J.
(Robert John), 1943–.
912′. 1338′0941   G1812.21.M1
ISBN 0-416-30290-4
ISBN 0-416-30300-5 (pbk)

Library of Congress Cataloging in Publication Data

Atlas of industrializing Britain 1780–1914.

Bibliography: p.
Includes index.
1. Great Britain – Economic conditions – Maps.
2. Industry – History – Maps. 3. Great Britain –
Industries. Langton, John. II. Morris, R.J.
(Robert John).
G1812.21.G1A2   1986   912′. 1330941   86-675435
ISBN 0-416-30290-4
ISBN 0-416-30300-5 (pbk)

Contents

List of maps

11   Textiles

23   Retail patterns

24   Labour protest 1780–1850

Acknowledgements

Those who believe that that which seems easy will be difficult and that that which seems difficult will prove impossible, would have had their expectations amply fulfilled by the experience of editing this atlas. The idea for the atlas arose as long ago as 1979 with the publication committees of the Economic History Society and the Historical Geography Research Group of the Institute of British Geographers. We were invited to become editors some months later. After many delays involving protracted negotiations with publishers, a research fellowship for one editor and a change of job for the other, we began to commission contributors in 1983. Discussions between the editors during this period brought considerable changes, expanding the scope and nature of the commission given us by the societies. We would like to acknowledge all those on the two committees who have given us help and encouragement. John Harris and Peter Mathias in particular gave help in different ways at different times. We hope that the resulting atlas will compensate those who expressed a little alarm over some of our proceedings. For us, the editorship has been a splendid experience in the relationships of pre-industrial production as we, the putting-out masters, negotiated between contributors who had all the ethic of craft producers controlling the work process, and the merchant princes of the societies. Anna Fedden of Methuen never lost faith in the project and deserves special mention.

The resource demands of the atlas have been considerable, especially as we needed to compile and draw many new maps and to co-ordinate thirty contributors. A timely grant from the Nuffield Foundation was vital to the progress of the cartography. The atlas would have been impossible without the resources of our two departments and of St John's College, Oxford. Such contributions are of course a normal part of academic life, but they need special mention at a time when harassment from state and commercial interests through cuts, copyright campaigns and threatened taxes is making the conditions of intellectual production especially difficult. Special thanks go to Jayne Lewin who drew most of the maps, and to Angela Newman who helped her and drew the rest; to Alan Hodgkiss for advice on cartographic style and design both before and during the early stages of drawing, to Dilys Rose for help in preparing the text, and to Barbara Morris for help and advice especially in the final stages of text editing. Our greatest debt is to our contributors who gave many hours to the project despite a variety of work and personal pressures. They and their departments gave generously of time and resources. We remain supremely grateful that so few refused our initial invitation and that none ran away when he realized just how demanding the preparation and compiling of maps could be. The demands for a minimum of consistency in presentation as well as the relentless pressures of space and cost meant that many small changes have had to be made in the text and cartographic intentions of several contributors, including the editors. We hope that they feel the overall result has justified these liberties, even where a favourite phrase or cartographic symbol has fallen.

John Langton
R.J. Morris
March 1986

Introduction

John Langton and R.J. Morris

Aims and ideals

The initial demand for the atlas came from those who simply wanted a collection of maps to illustrate their teaching. We rapidly realized that here was an opportunity for something more than a cartographic anthology re-drawn to uniform conventions. Amongst those involved in the two societies were a number of people whose knowledge would enable us to map the spatial patterns and relationships of a wide range of human experience during the industrial revolution period in Britain. We began with the basic headings of 'economic history' as it stood in the 1970s, and looked for ways in which the broad concerns of that literature, with inputs of raw materials, labour and capital, with technologies of power and transport, and with the changing nature of output (Mathias and Postan, 1978), could be expressed cartographically. We also drew in those who were outlining and explaining the social aspects of this period of change.

This does not mean that the atlas pretends to be a theoretically coherent whole. Our choices were limited not only by the basic constraints of space and cost but by our sense of what was readily mappable given the current development of our subjects. The gaps remain as a challenge to those who have the time and ingenuity to fill them. More important, users of the atlas must not be led by the necessary ordering of material within the book to imagine that the editors conceived a rigid theoretical plan. By placing environment and population first we are not suggesting a cause–effect relationship with the patterns of economic activity which follow. Nor by placing sport, language and religion last do we intend to reduce these to the status of mere epiphenomena. Indeed the major lesson of laying out such a variety of human activity spatially has been to suggest to us that the pressures of causation move in many directions. We hope that one result of mapping this variety of human activity ranging from manufacturing and wealth accumulation to sport and dialect publishing will be to counter the regrettable tendency for economic and socially orientated concerns to become separated in current academic discourse. Although transport and settlement patterns have a greater place here than in most economic histories it is fair to say that the initial ordering of the material depended very little on theories with a substantial spatial element in them. This is important for the conclusions implied by the maps clearly do have an important spatial content.

We determined that the atlas should refer to the industrial revolution in Britain, that is England, Wales and Scotland. Despite its importance Ireland presented problems of data availability and compatibility and indeed involved key processes and directions of change that were fundamentally different, so that we felt it should properly be left to historians of that island. The insistence on including Scotland caused enough problems of data acquisition and standardization especially for English contributors. As ever, the Scots found it easier to deal with English data. Even so, the different legal and institutional frameworks of the two parts of Britain mean that map users must be careful to allow for the different basis in time and practice between series from Scotland and from England and Wales. In a few cases we have inserted a small guttering at the border as a reminder. The poor law and the religious census were two cases where national differences influenced the collection of data.

We emphasized to our contributors that we conceived the 'industrial revolution' not as an event but as a process operating over a considerable period of time, and operating in different ways at different times and places in a variety of aspects of human activity.

Current literature no longer makes it possible to see the industrial revolution as a cluster of innovations, or a break in long-run national economic trends sharply defined in time (Crafts, 1985). The maps, like the basic statistical series, show that the economic and social structure of Britain was very different at the beginning and end of the 19th century. It is clearly nonsense to argue over how far it was cotton spinning mills in the 1790s, or intercity railways in the 1840s, or even the machine tools of the 1860s which have prior claim to our attention. Perhaps some only felt the impact of the industrial revolution when turnstiles were fitted to the enclosure at major horse-race meetings. Although all contributors see the 'industrial revolution' as a process which involved massive changes in the organization of production, the productivity of economic processes and the structure of output, we have not sought to impose any sharper definition than that. Hence the wording of the title of the atlas. Divergencies of views between contributors will be evident to all who use the atlas. At least one contributor sees the process as one involving not so much industrialization in the manner of the traditional historiography as a shift from agricultural production to services.

We do not want the atlas to become a resource which is plundered by proponents of different interpretations for evidence to support their viewpoint. We hope that the collection will persuade all schools of thought to examine the spatial aspects of the processes and changes which they hold to be important. In many cases they will find that changes in spatial organization, such as the distribution of industries, transportation, population and wages, preceded the major changes in output, technology and productivity which have attracted most attention, at least from the historians. Most versions of what the 'industrial revolution' was can be found in the atlas in some particular phenomenon, region or span of time. This should not be disconcerting. The spatial and temporal deconstruction which the atlas provides is an important part of understanding the nature of industrialization.

Data and the critical use of maps

Contributors and editors alike have found the demands of cartography exacting. The map is greedy for comprehensive and consistent data in a more total manner than any tabulation or regression. In a table missing data require an extra line or can be handled by reducing the total population dealt with by the table. On the map missing data stare out as a blank, often risking giving a misleading impression. We rejected the idea of maps which simply located economic activity in a general way from impressionistic accounts – the 'here be dragons' variety of map. We sought maps with reasonably consistent, quantifiable and complete area or point data. In this form maps probably represent a greater density of readily accessible information than any other form of exposition. Having made this demand we were impressed by how late it was that Britain entered statistically observable time in most areas of human activity. For many topics little was available before the 1820s. There were exceptions. Foreign trade was one, but despite the enormous amount of work done on the population history of the eighteenth century, there is not enough yet to attempt anything like a full spatial representation of the changes in light of the recent revisions of this population history (Wrigley and Schofield, 1981). Data availability pushed the atlas forward in time. Now if we see the industrial revolution as a process then this is no bad thing. We know from the work of von Tunzelmann (1978) on coal and steam engines, Hawke (1970) on railways and Hyde

(1977) on iron smelting, that major changes in productivity and production did not come in many industries and regions until the 1840s. Still, it is very clear from the information which we have got that some important spatial changes came much earlier, as the maps of wages and of the chemicals, textiles and iron industries indicate. There were many important spatial changes in the eighteenth century which are as yet half hidden. Deane and Cole (1962) have suggested that high birth rates were associated with eighteenth-century Lancashire in particular. Corfield (1982) has shown that the beginnings of the replacement of one urban system by another also took place in the eighteenth century. Most of the dense patches of population evident in 1801 must have emerged in the eighteenth century. In many sections of the atlas, the earliest maps show that much had been done to erode London's primacy since the late seventeenth century (Wrigley, 1967). Important matters, such as the control and possible changes in the control of production by the finance and institutions of London can only be outlined in a general way. Perhaps some brave soul will attempt an eighteenth-century atlas. The spatial changes between 1820 and 1914 proved important and interesting enough mainly to occupy one volume.

The lines and shadings on the maps seem hard and authoritative, but those who read the text will find that many authors experienced a continual struggle with the consistency and meaning of their data. Warnings abound. The data on which these maps are based were created in a variety of ways for a variety of reasons and survive in sources which vary from census, taxe and denominational records to plans, bibliographies, diaries and parliamentary surveys. As with all historical sources and their manipulations by historians, the results need to be interpreted critically. In places data gathering was incomplete and lacked precision. More important, the categories used for data collection may not suit modern analytical needs. We would like to warn two groups of readers. Some will be tempted to accept the information on the map in a direct and immediate way. Others, aware of the complexity of data creation and the occasional arbitrariness of categorization may feel that the maps are so tied about with qualifications as to be valueless. We hope that the atlas will create amongst all its users a continual awareness of the need for the careful and critical use of data and their representation. Here are a few general guides which apply to the atlas.

(i) In general, contributors have used variables with a direct and easily comprehensible meaning, such as crude death rate rather than a standardized rate. Although this is helpful it should not be allowed to conceal the complexity of the situations which created these variables and their values. Thus a high death rate is often attributable to the age structure of an area rather than to its inherent unhealthiness. In nineteenth-century conditions of high infant mortality a very young population usually produced a higher crude death rate, other things being equal, than a population with more people in middle and old age.

(ii) The proliferation of point and area data on the maps makes this atlas a supreme temptation to the ecological fallacy. Remember that with very few exceptions the variables record the characteristics of individuals and organizations within the areas and not characteristics of the areas themselves. The people who possessed one characteristic in an area were not necessarily the same as those who displayed another. Thus, although the distribution patterns of many variables have much in common, as we discuss below, those variables must not be related to each other simply because of that common pattern. There is a fair coincidence of labour disputes and wealth accumulation, of professional football and co-ops, but without a much deeper enquiry not even a hint of a cause–effect relationship should be acknowledged.

(iii) These and other difficulties may sometimes be thought through in terms of intervening variables, which are often hidden from the first glance at the maps. The age structure of the population was clearly such a variable behind birth and death rate distributions. The matter of sport and co-ops almost certainly had no direct relationship, but a more general feature which we might call working-class culture as a convenient shorthand probably linked them together.

(iv) There is a special case of an intervening variable of a technical nature which needs mention in this atlas, namely size of area. The bigger the areas used in spatial classification, the greater the probability that smaller scale variations are being averaged out. The bigger the variety of size amongst the areas in the map base, then the bigger the chance that this is being done to a different extent over some parts of the map compared to others. This also affects the picture shown by maps of absolute values.

The large northern counties were a problem in this respect. In some cases disaggregation of the Ridings of Yorkshire helped. In others, dividing variable totals by population totals yielded an index which was scale free. Other contributors like Overton have used proportional indexes. As the maps rarely show absolute figures, it is important that readers note the form of representation before drawing conclusions and making comparisons between the different maps. The effect of area size and its variability should always be critically examined. For example Overton's maps show a ring of market gardens around London. The presence of this ring on the maps is in part due to the reality of London's economic influence, but it is also due to the small area of the counties around London. Spatial disaggregation might have produced similar effects around other large towns. Even without it, there is a hint of the same impact on Cheshire from the south Lancashire towns.

(v) The concept-indicator problem is an important part of the way in which users and contributors respond to the maps. Historians and social scientists use information created by past administrative and statistical processes to represent concepts and ideas required by their own theoretical and analytical curiosity. In many cases the gap between concept and indicator was small. Death rates are in fact deaths recorded by the Registrar General. The only problem here is a small and unknown amount of under-recording which had a disproportionate impact on the very young. This sort of gap is more problematic in the case of religious adherence. The census, as McLeod points out, in fact recorded attendances. Thus the figures are plagued by under-recording and double counting. This would be fine if the relationship between adherents and attendances was constant, but we know that double and treble attendances affected nonconformists more than established churches and that non-returns were greater in Scotland. Another form of concept-indicator jump is made when a part is made to represent the whole. Thus Hunt firmly presents his wage data as that for carpenters and agricultural labourers in specific locations, but his discussion suggests that much broader and more interesting historical statements can be made if we cautiously assume that the relationships shown hold for regions and the wage structure as a whole. There is nothing wrong with moving from the safe to the speculative level and using the part to represent the whole. It is important, however, to be aware of the risks and intellectual decisions taken in making such a move.

(vi) The maps usually present one variable distributed in a two-dimensional space. This produces a temptation to seek one-factor or uni-causal explanations. These can often be confusing and misleading. Look at Lee's maps of indoor domestic servants in 1851 and 1911. A naïve interpretation of domestic servants as members of middle-class households might be worried by the distribution in agricultural counties. Once it is realized that domestic servants included large numbers of girls who not only helped in the house, but also worked in the shop and above all helped with light farming tasks, feeding chickens or working in the dairy, then a double pattern emerges dominated by the service economy of the south and the agricultural counties, especially those in the cheese and dairy areas. The overlapping of two or more patterns on the one map is very common.

Undisclosed spaces

If we match an 'ideal' list of contents to the atlas there are many omissions. Some were simple matters of choice within a tight space budget; leather not non-ferrous metals, brewing not fishing. We hope that these limitations enable our publishers to sell at a civilized price. Other gaps deserve more extended comment. In many ways, the atlas represents the 'state of the art' for the two disciplines. We hope that dissatisfactions with the resulting balance will encourage more research on spatial change. One area requiring more extended research is clearly the eighteenth century, which does not have the readily available series from parliamentary papers as does the nineteenth. For the post-1820 period the atlas presents a clear picture of distributions at various points in time, but, with the exception of transport and migration, there is little on flows. Point-of-production information is only the beginning of explanation. Information on sources of raw materials and destinations of the product is needed to answer many of the questions posed by the maps, but information on flows of all kinds is difficult to come by. Lack of space and the technical difficulties of superimposing patterns at different times on one map have meant that change over time has got less systematic attention than it needed. The population and urban sections are ones which show that the distribution given in absolute quantitative terms is very different from that indicated by rates of change.

Other, perhaps more challenging, gaps come from a lack of conceptualization and investigation in spatial terms or at a scale appropriate to the atlas. Gender relationships get very little attention in the atlas. In part this is because gender theory has been at its weakest when dealing with large-scale spatial relationships. Recent moves in this direction came since the atlas was conceived (Women and Geography Study Group of the IBG, 1984). This is in sharp contrast to several powerful analytical contributions regarding gender differences in the use of space within the house and neighbourhood (Ardener, 1981). Regional and local variations in sex ratios mapped here suggest that the links of gender to labour markets and family needs in the context of migration need fuller exploration, perhaps in the former case along lines sketched by Massey (1984).

Our understanding of capital – its availability, its flows, its cost – is likewise difficult to organize in spatial terms. Neo-classical assumptions hold even greater sway here than in labour theory. Only recently has capital been conceptualized in spatial terms (Harvey, 1982). What should be mapped? It is the growing institutional structure for mobilizing capital which is presented here. Mapping rates of interest would in theory produce uniformity, but did liquidity and credit availability have a spatial aspect that was not demand-generated and how efficient were the mechanisms linking spatial variations in demand for and supply of capital? It may be that credit was easier in an urban context and that the real gradient was between the industrial village and the urban workshop. In any case, there must have been larger-scale variations in rates of fixed-capital formation and the supply of capital finance, and hence strong flows of capital over space. Economies do not really exist on the heads of pins. But there are simply few data sets to map except those devised by Cottrell to describe the institutions formed to cope with these geographical realities.

Spatial patterns revealed

The major achievement of the atlas and its contributors undoubtedly lies in the period 1820–1914. Here the maps taken as a whole reveal a number of competing sets of spatial patterns which are worth drawing out, even though doing so involves some speculative suggestions about the earlier period. These sets of patterns were all part of the reality of economic and social change in the period. The tensions and conflict between these patterns were aspects of reality not invitations to either/or debates amongst intellectuals. We need to tease out which variables show which set of spatial relationships and how these change over time, and not to impose preconceived models on anything and everything revealed by the maps.

The complexity of the patterns revealed by different maps, let alone those revealed by combining different ones, are very great, demonstrating how strongly varied the lived experiences of people must have been from place to place. This variation was not chaotic. The complexity stems from the simultaneous operation and changing relative intensity of a number of spatial processes – that is, historical processes which produced intelligibly ordered geographical patterns. Three can be distinguished on the maps, although they often worked to confound and complement each other. First, there was a continuing core–periphery relationship between metropolitan and provincial Britain. Then, there was the pattern of the resource-based industrial development of areas rich in water power, coal and other minerals, which lay along the junction of the highland and lowland zones. Other processes were always operating at a smaller scale to link the way in which people think, act and organize themselves, through their work, daily routines and relationships, to the resources of their immediate environment.

The maps show clearly that despite the customary emphasis in the literature on the coalfield regions, Britain was and remained a metropolitan economy. London was the largest city. Its citizens controlled the greatest accumulations of wealth. It dominated the major part of the banking, commercial and political systems. The maps of the railway system show that London and not Lancashire was the centre of the network. That being said, the resource-based regions of Yorkshire, Lancashire and the West Midlands emerge as major areas of social and economic innovation. Population and urban growth rates were highest here. The textile, chemical and engineering industries cluster here, especially before 1851. Lancashire and Yorkshire emerge as distinctive areas of cultural innovation within the group. Chartism, co-operative societies, other popular self-help organizations, dialect publishing, and professional sport all pick out this trans-Pennine region. After mid-century, south Wales and northeast England take a fuller part in the resource-based economy showing their own distinctive patterns of growth and cultural change. Their later surge was associated with a more complete dependence on large-scale heavy industry. The coal and iron industries in these regions did not generate local spin-offs into secondary and tertiary activities to the same extent as had occurred in the earlier period elsewhere. Perhaps this was because of a more 'colonial' relationship with metropolitan England (Hechter, 1975).

The spatial divisions thus produced have a striking resemblance to the highland/lowland division of traditional geography which is so evident in geology, geomorphology and climate (Mackinder, 1907; Hodgen, 1952). Lee's analysis of economic structure provides an important reformulation of this division, as a divide between the commercial and service economy of the south and the industrial and extractive economy of the north and west. This process served to reinforce the differences in economic organization and culture which had always been strongly evident in agricultural Britain (Snell, 1985). These still showed through in the maps of nineteenth-century patterns, not only in the pastoral/arable production difference, but also in rural population structures and growth rates, poor-relief expenditures, religious affiliations and farm-labour organization. Indeed work on proto-industrialization and the development of agrarian capitalism in an earlier period suggests that these cultural differences were important stimuli of the different rates and styles of economic development (Langton and Hoppe, 1983). They were not null relics of long-defunct processes. The tension between the metropolitan core–periphery structure and the resource-based structure is evident in many maps. Hunt's analysis of wages identifies growth poles in the resource-based regions, but London still remained the high-wage area. The maps of trade societies and friendly societies show London-based clusters as well as Yorkshire and Lancashire ones. The urban growth rate distributions show not only the new industrial towns of the north and Midlands but also the ring of resort and suburban towns around London. The importance of London was sustained by its position as the capital city of a growing world empire, by the increasing size and complexity of central government in Great Britain itself, as well as by the continued prosperity and importance of London's port, financial institutions and specialist industries. The rise of the provinces in political and economic power was a dramatic reality but only in a relative sense. The mills and co-ops of Lancashire and Yorkshire could challenge but never dominate the banks, public schools and broad acres of the south. The north could carry away the FA Cup and dominate historians' league tables for rates of change and growth but could never dominate in terms of accumulations of wealth and population.

There is evidence that by 1900 the metropolitan economy was beginning to reassert itself against the earlier relative losses to the resource-based regions. Its primacy, never completely lost, began to increase again and indicators of economic power like stock exchange membership and tax yields showed resurgent dominance. At the same time, the resource-based economies of the provinces were becoming more completely based upon industry. As the maps of textiles illustrate, these areas were becoming more narrowly dependent upon traditional industrial specialisms. The multiplier effects generated by the industrial growth poles of the provinces were leaking out to metropolitan England in the more integrated national economy of the late nineteenth century. This added to the influences of imperial nodality and centralized national government on the one hand, and of highly capitalized agriculture and local growth industries on the other. These processes eventually fused and produced the stark contrasts of the inter-war years (Massey, 1984; Dunford and Perrons, 1983).

In a more difficult way the related question of regionality is raised: just how self-contained and separate were the various industrial regions, even during the earlier phase of most rapid relative growth? This is clearest in the transport material. The flows of goods along the canals and turnpikes of Lancashire and Yorkshire are clearly greater than flows out of the region, except for the export funnels of Liverpool and the Aire.

Earlier the long-run influences in economic control moved from the national system of the Blackwell Hall factors and London ironmongers to the regional Cloth Halls and warehouses. In the nineteenth century, regionally based pressure groups became highly influential in national political life (Read, 1964). Do these trends represent the growth of integrated and discrete regional economies? The canals, tied to topography more than any other transport system, denied the central place to London which the east coast trading system had encouraged. It was during the period when canal haulage articulated the industrial economy that the large commercial cities of the provinces grew most rapidly in relative and absolute terms (Langton, 1984). The railways went some way to reasserting London's central place. The railways were a dense and topologically highly focused network through which the forces of national economic integration very evident in the late nineteenth century worked in powerful ways. Even so, the regional base of many companies meant that pricing policies encouraged intra-regional trade, quite unlike the 'straight through' fares policy of railways in India and Canada which discriminated in favour of national and export trade patterns. One railway historian has likened the effects of early railway company fares policies to those of regional development agencies (Hawke, 1970).

It is in the matter of regionality and the way in which it was changing, as in the question of the relationships between Lee's two economies or between core and periphery, that we need greater knowledge of flows of capital, raw materials, finished and semi-finished goods and cultural and political influences. Attention needs to be concentrated on patterns of railway shareholding, stock exchange transactions, company amalgamations and shifts in their head offices and the spatial patterns in government revenue gathering and spending.

Britain, as ever, was in the nineteenth century, a nation of nations. National identities remained in a variety of ways. Wales was marked by language and a distinctive religious response to industrialization. Some of its distinctive nature was a result of older cultural characteristics. Other aspects were part of the wider pattern of the highland/lowland distinctions in farming, and some a result of the particular nature of its industrial resources and their exploitation by capital based outside the principality. Although Scotland had lost independent statehood a century before the atlas comes into focus, its distinctive nature still remained within the creation of the British economy. Edinburgh and the Lothians showed in a miniature way many features of the London and southeast England service-based economy. Many cultural patterns were distinct. The border was indicated by marriage patterns, religious practice and housing provision. In other matters, like the poor law, institutional structures were distinct, but the results of their activities show a convergence of practice and experience with the English pattern. In many ways these national patterns were survivals from the eighteenth century, but we need more evidence before we can be sure whether these patterns were only dissolving anachronisms or were national patterns re-forming in new ways as dynamic elements in the processess of nineteenth-century social and economic change.

This is even more true of the smaller-scale regional distinctions within each of the nations. The maps show them as complex and indistinct. Existing literature gives little help in making sense of what the maps show. Work which will provide us with a picture of the regional structure of pre-industrial England, like that provided for France by the *Annales* school of historians, is really only just under way (Everitt, 1979; Thirsk, 1984;

Underdown, 1985). There is even less for the nineteenth century itself (Joyce, 1980; Calhoun, 1982). The countrywide data base used by most contributors to the atlas is often too coarse to reveal many regional differences that did exist, either as survivals from an earlier period or as responses to varied industrial experiences. Even so the maps do show clear differences between the economic bases and cultures of broadly defined regions within the north and Midlands. The point-symbol maps of religious affiliations and political preferences give them sharper resolution, as do the maps drawn at a sub-national scale. There are hints of smaller-scale distinctions between the four northern-most counties and the rest of northern England, between southeast and southwest Lancashire, between the northern and southern parts of the West Riding of Yorkshire, between the east and west Midlands, and between the champion and wood-pasture regions in the south. We have little knowledge yet as to how far these differences produced distinct, integrated and coherent regions, nor how far they were a result of historical survival or developing industrialization.

This discussion has shown that it would be quite wrong to formulate starkly polarized either/or questions. Whether cultural inheritance or economic development is construed to be an independent element will depend very much on the time-scale within which the historical process is conceived. It is obvious that they interact to change one another. There was no cause and effect but a dialectical process of reciprocation. Neither of these elements was spaceless. Different historical processes have different spatial logics, and the patterns they produce interact with each other. This dialectic of spatial processes is a result of the fact that they are geographical expressions of historical processes. Each of the sets of patterns we have isolated was to an extent dependent upon the others for its shape and dynamic. The core–periphery relationships that integrated British economy and society produced one sort of large-scale pattern, but they also intensified smaller-scale regional distinctions through encouraging specialization and competition. The provincial industrial regions were at first nourished through links to London which brought finance, techniques and access to wider markets. These links were never broken. Eventually they took back, through multiplier leakages, resources which strengthened London's position. The dynamism and variety of the parts increased at the same time that they were being moulded into a wider core–periphery system. The core grew because the periphery was becoming more dynamic. In culture and politics, as well as in economic terms, dependence of one kind flowed from independence of another.

What the maps do show is the complexity of competing and complementary spatial patterns in a period of sustained economic growth and social and political development. We hope that the atlas as a whole will be used as a basis for thought and argument about the processes that created geographical differences in Britain, and about the significance of those differences for the history of the period, as well as providing illustrations of the spatial distribution of particular economic and social phenomena. Published in a decade when power and prosperity threaten to concentrate in one area of the kingdom, the atlas is a reminder of the manner in which prosperity and vitality depended on the variety of place and region, and on the tension between different patterns of spatial organization. In current circumstances increasing awareness and understanding of these different patterns, whether in teaching or research, can only be a benefit.

Atlas of
Industrializing Britain
1780–1914

1 The physical environment

John Langton

The physical environment of Britain is immensely varied, probably more so than that of any other area of a similar size on Earth. This variety is contained within a general contrast between highland and lowland zones, separated along a line running from the Exe to the Tees estuaries (Stamp, 1946).

The complete range of sedimentary strata from the earliest Pre-Cambrian underlie Britain, some of them metamorphosed or intruded by igneous rock. Folding, faulting and long periods of erosion have brought all of them to the surface (1.1). The sequence of outcrops runs from the north and west to the south and east, severely broken only by the great faults which bound the central lowlands of Scotland.

Generally, the lowest schist sediments, metamorphic and igneous rocks are hardest and therefore most resistant to weathering and niggardly of soil, often giving rise to steep slopes and pointed crests. The slates, shales and sandstones of the Lower Palaeozoic are also hard, but except in the vicinity of intrusive lavas in north Wales, the Lake District and southwest Scotland they normally yield more rounded contours (1.1 and 1.5). The Upper Palaeozoic strata include the hard sandstone known as Millstone Grit which outcrops along the Pennine anticline; the Carboniferous Limestone which, owing to its permeability and consequent resistance, rises above it in the Peak District and north Wales, and the later sandstones and shales interbedded with coal seams known as the Coalmeasures, which floor much of the Scottish lowland trough and outcrop in an elongated broken arc around the Pennines, from Cumberland to Warwickshire to Northumberland. The alternations of shale, sandstone and resistant limestone beds in the Carboniferous and of soft sandstone and limestone in the Permian rocks deposited on top of them have created a complex topographical transition between highland and lowland zones (1.5).

The more recent rocks of the lowlands to the southeast of this transition are generally softer and less affected by tectonic activity. However, the alternating outcrops of porous limestone, sandstone and chalk separated by soft shales and clays give considerable topographical variety. The main cuestas are formed in the Jurassic outcrop which extends from the North York Moors to Dorset, with limestone and sandstone scarps between vales of shales or clay. These formations are separated by the wide vale formed on Oxford Clay from the chalk outcrop that stretches from the Yorkshire Wolds to Salisbury Plain. Large areas of the Jurassic scarps are over 350 feet high, whilst the vales are low-lying and were subject to severe inundation near the coast, creating extensive wetlands (1.5). The southern ends of these Mesozoic beds were compressed in Dorset and bunched around the Wealden anticline in Sussex and Kent by the Alpine orogeny, hence the considerable variety of relief in those areas. The most recent and softest Tertiary rocks cover a relatively small area in the embayments around the Solent and Thames estuary.

Much high land was planed by ice, but the hard lavas and granites of the highest land of all in north Wales, the Lake District and the Scottish Highlands were deeply etched by corries and valley glaciers. Thus, even the relatively low highland areas of Britain are often fully 'mountain-like' in appearance. Flattened upland surfaces were plastered with boulder clay by melting ice, and blanket peat has further enveloped them. Glaciated lowlands were also covered by boulder clay, whilst deposition in ice-dammed lakes, by meltwater streams, solifluction and winds blowing off the glaciers smothered large unglaciated areas with sand, gravel and brickearth (Sparks and West, 1972). Later

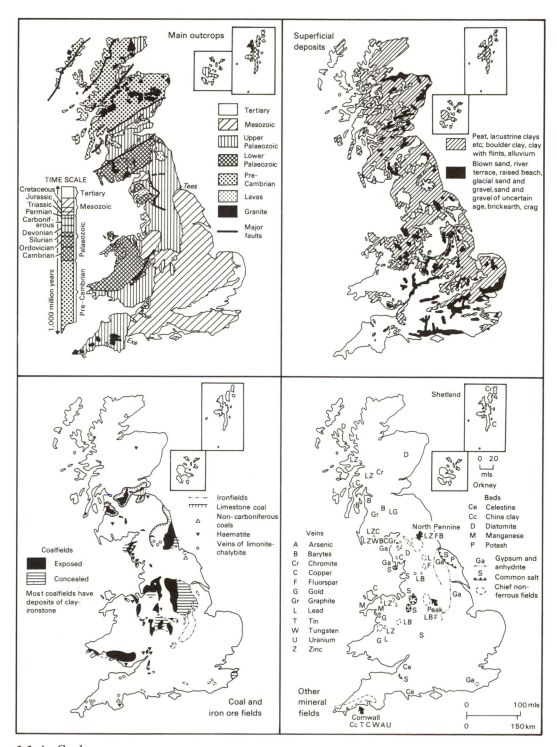

fluvial and marine deposition covered the lowest-lying land with alluvium and silt. Thus, most of Britain's land surface is not derived directly from the geological beds immediately beneath (1.2) and soils are therefore even more varied from place to place than lithology alone would suggest (Coppock, 1971; Curtis *et al.*, 1976).

The Carboniferous, Permian and Jurassic beds at the junction of highland and lowland zones are particularly rich in minerals (Eastwood, 1964). Coalmeasures outcrop over large areas of the Upper Palaeozoic beds (1.3). Up to forty seams contain many types of coal. Workable coals are also present in the Millstone Grit and Carboniferous Limestone beds below the Coalmeasures themselves, and spread beneath Mesozoic strata as concealed coalfields, especially to the east and south of the Pennines where faulting was less intense than to the west or on the margin of the Welsh Massif. Coalmeasures even rise to shallow depths beneath later strata in western Oxfordshire (discovered in 1877) and Kent (discovered in 1895) (Hull, 1905). The scattered pockets of coal in later strata were locally important in the nineteenth century. The seat earths at the base of many coal seams, especially the lower ones, and the highly aluminous clay produced by the weathering of Millstone Grit lava in central Scotland, provide the raw material for refractory and sanitary wares; Etruria Marl for hard blue engineering bricks. Ironstone bands and nodules in Coalmeasure shales, particularly abundant in Scotland, west Yorkshire, Staffordshire and south Wales, supplied the bulk of the iron ore used in Britain until the second half of the nineteenth century. Even higher quality haematite ore was present in the Carboniferous Limestone in south and north Wales, the Forest of Dean, Furness and Cumberland, whilst the lodes of iron ore that occur as siderite or chalybite in the same strata in the Pennines, Devon and Cornwall and the pockets of haematite and other iron ores in western Scotland gave rise to brief but frantic mining activity in the nineteenth century. After the exhaustion of the Carboniferous ores the majority of home-produced iron ore was raised from the lower quality Jurassic deposits, which are particularly abundant in Cleveland, north Lincolnshire, Northamptonshire and north Oxfordshire.

Other non-Carboniferous minerals were heavily exploited by early industry. Primary amongst them was Cornish tin, used in pewter making and plating in south Wales. The copper of Devon, Cornwall and Anglesey was also mainly shipped out for processing on coalfields; so was the lead of north and central Wales and the Pennines. The even rarer metal ores plotted on Map 1.4 were only ever mined sporadically or on a small scale. Large deposits of bulkier lower-value minerals, on the other hand, often supported the growth of heavy industry and rapid urbanization, especially if they were located on or near to coalfields. The salt, potash and gypsum of the Permian beds in Cheshire and north Yorkshire gave rise to important chemical industries, and the potteries of north Staffordshire were based on local clay and china clay imported from the extensive deposits of Cornwall, as well as coal. Town building made heavy demands on mineral resources. The high-quality slates of north Wales, Westmorland, Aberfoyle and Ballachulish roofed many of the towns, tiles made from high-quality Coalmeasure clay the others. Chalk was quarried for cement, blown sands – especially the white non-ferrous Shirdley Hill sand of the St Helens area – for window glass. The wide variety of clays used in brickmaking gives some pointed visual contrasts among urban townscapes, from red Accrington to white Cambridge, but the Oxford Clay, which contains almost enough bituminous material for its own firing, was particularly heavily used, especially in the Bedford and Peterborough areas.

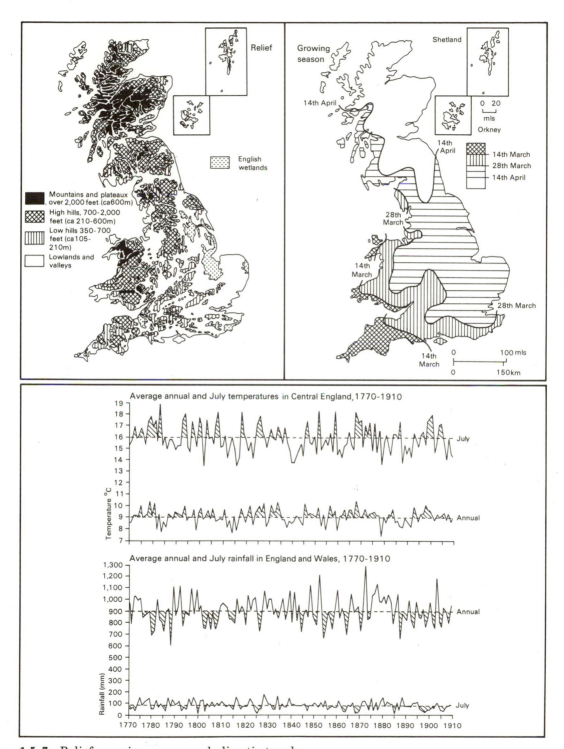

1.5–7 Relief, growing season and climatic trends

Because of the Gulf Stream and the southwesterly origin of much of the air that passes over the island, Britain is much more temperate than its northerly latitude implies. This predominant air flow is often in contention with polar air or air masses based on the continent. This gives enormous day-to-day and year-to-year unpredictability, but usually within a narrow band of extremes (1.7), so that the British climate is 'the most dependable in the world' (Manley, 1952, xiii; Manley, 1974 and Wigley *et al.*, 1984).

Transgression of the normal limits to these fluctuations, as in 1792 and 1816, could cause severe harvest failure, but such extremities are notable for their rarity. Economic significance has recently been claimed for the shallow trends in the series on 1.7: the cold winters of the early nineteenth century in impeding canal haulage (Freeman, 1980); the increased variability of rainfall in the West Riding of Yorkshire in the 1820s and 1830s in making water power unpredictable and encouraging the diffusion of steam engines (Gregory, 1982); the growing coldness of the late nineteenth century in lowering the limit of cultivation in the uplands and contributing to the Great Depression of arable farming (Parry, 1978). However, the effects of such shallow average changes (and of occasional extremes) differ widely from one part of Britain to another (Doornkamp and Gregory, 1980) because of the sheer variety of the British climate from place to place.

Maritime influences keep winter sea-level temperatures high for the latitude everywhere in Britain, but most markedly so in the west (1.8). It should be noted, however, that the effect of altitude will tend to lower actual surface temperatures further in the west – by about 3°F per 1000 feet – than the east. In the summer there is a much more pronounced south to north progression, though still some coastal moderation (1.9). In consequence, the western half of the island is, at any given altitude, much more equable than the east: the average dates of first and last frosts are almost the same (15 November and 1 April) in the Hebrides and Kent; winters are harder, with more frequent frost and snow, in Surrey than the Orkneys, and the growing season begins sooner in Cumbria than East Anglia (1.6) (Bilham, 1938; Chandler and Gregory, 1976). Unlike perennial grasses, annual crops such as cereals and roots are more affected by the intensity of the summer heat for maximum growth and ripening than by the length of the growing season (Coppock, 1971). The map of accumulated day-degrees (the annual cumulative total of degrees by which average daily temperatures exceed the minimum necessary for plant growth) shows the net effect of the much higher summer temperatures of the south and east despite later springs and earlier autumns (1.10).

More than 100 inches of rain fall annually on the western mountains and a large part of the western side of the island receives over 40 inches a year (1.11). In these areas, rain falls on more than 200 days a year, 250 in the wettest parts of Scotland (Bilham, 1938). Parts of eastern England, rainshadowed by the western highlands and more influenced by continental anticyclones, have less than 25 inches of rain per year. In parts of this drier zone the monthly maximum of rainfall comes from summer convectional storms, and thunderstorms are much more common in the south and east (1.12). However (unlike in France, where they were more feared by the peasantry than any other climatic hazard) these summer thunderstorms are rarely accompanied by hail, which is most frequent throughout Britain in March, before crops are showing but when bare and already soaked soil can lead to rapid runoff and severe flooding in low-lying areas such as the Fens.

The heavy rain of the highland zone causes leaching and acid soils. The cloudiness

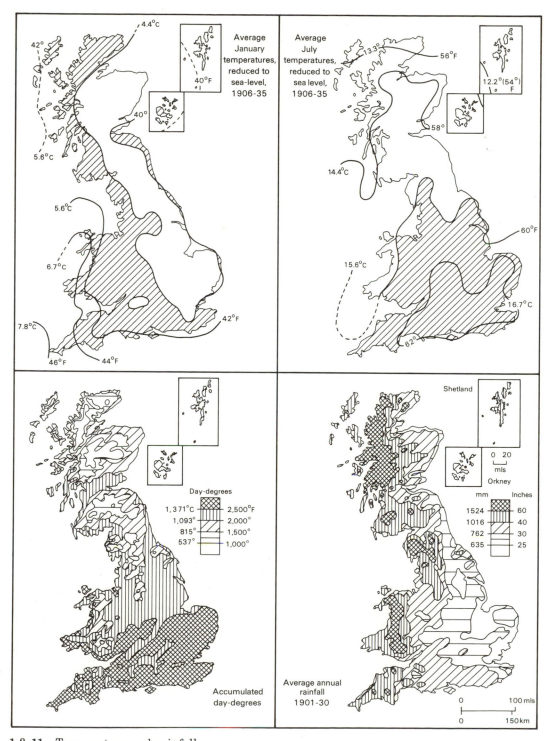

1.8–11 Temperature and rainfall

and lower temperatures make the ripening of cereal crops less certain than in the south and east. Crop failures in Britain are normally due to cold wet summers and consequently they are more common and more severe in the highland than in the lowland zone. The highland zone is extremely well suited to grass growth; pastures can survive through summer despite their shallow roots and continue to grow through frostless winters. Only waterlogging prevents animals overwintering out of doors in the western valleys and coastal areas. In the east, grass can only survive summer in the dampest valley bottoms, but deep-rooted cereals with high summer temperature requirements thrive; dry winters allow cultivation when it is needed to maximize yields and heavy frosts break up turned soil and kill weeds. A very strong climatic contrast thus lay at the root of the deep difference between the pastoral and arable agricultural regimes, and therefore the human cultures, of pre-industrial highland and lowland Britain (Fox, 1959; Mackinder, 1907).

The largest deviations from normal are most likely in the southwest Midlands, not in the extreme rainfall zone of the southeast (1.13). Over most of the island rainfall of within 15 per cent of average can confidently be expected. However, the combination of low rainfall with a summer peak and high summer temperature means that soils so dry out throughout southern England that irrigation is necessary to maximize production in at least 5 years in 10, and in up to 9 years in 10 in the extreme southeast (1.14) (Coppock, 1971). The sand- and chalk-surfaced areas of the lowlands were therefore barren, usually covered by heath, forests or scrub, until the introduction of four-course rotations allowed large quantities of moisture-retaining manure to be applied to them. Conversely, the wetness of the lowland clays in winter and spring prevented the heavy cultivation necessary to maximize yields until underground drains were laid in the nineteenth century.

The abundant rain of the west does not flow through river systems to the summer-parched eastern lowlands. The eastward-draining rivers of the south are all contained in catchments with low runoff totals caused by low rainfall, high temperatures and porous rocks (1.15). Thus there has always been a shortage of surface water for irrigation and summer domestic use in the lowland zone. Since the middle of the nineteenth century this had been compensated by pumping through boreholes in the chalk and sandstone aquifers that previously yielded only well water. The water table in the chalk below London was lowered by 75 feet by 1877 and a further 150 feet by 1927 (Walters, 1936; Sheail, 1982). In highland Britain a much higher proportion of the much heavier rainfall flows into a far denser network of rivers (Ward, 1981). River discharges are considerably greater: the Scottish and Welsh Dees carry ten times as much water as East Anglian rivers at all times of the year and their lowest summer flows are four times the winter maxima of East Anglia. Moreover, the generally rugged topography means that all the rivers of the highland zone are 'youthful', with steep channels and side slopes, little floodplain and low levels of suspended sediment load (Newson, 1981). Water power and pure water were thus in enormous abundance almost everywhere in the highland zone in the early stages of industrialization. Later, the same geomorphological and hydrological characteristics provided ideal conditions for urban water supply.

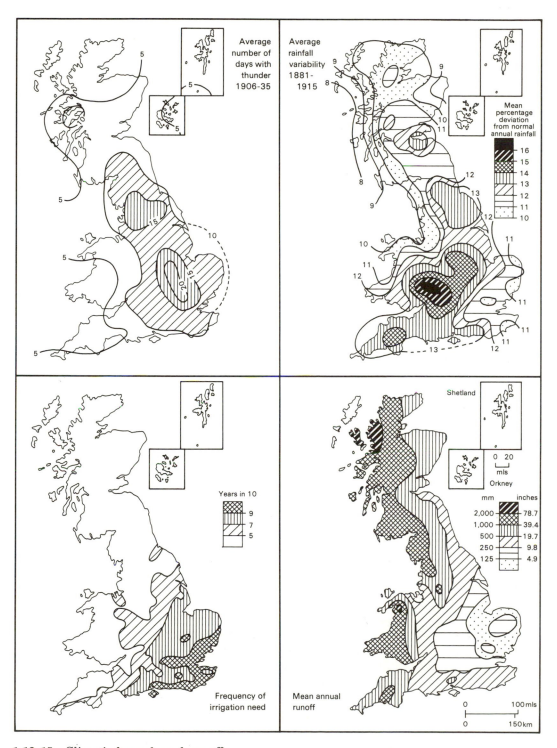

1.12–15 Climatic hazards and runoff

2 **Population**

Richard Lawton

Information on the numbers and distribution of the British people is very incomplete before the first census of 1801. Sporadic English parish listings, and Webster's survey of 1755 and the Old Statistical Account of 1791–9 for Scotland provide useful landmarks (Flinn *et al.*, 1977), but the best population estimates are derived by back-projection from censuses using parish registers (Wrigley and Schofield, 1981). The lack of migration data makes these techniques unsuited to regional study.

The population of England and Wales increased from 6.0 to 8.9m between 1741 and 1801, doubled to 17.9m in 1851 and doubled again to 36.1m by 1911. Scotland's population grew more slowly from 1.2m in 1755 to 1.6m in 1801, 2.9m in 1851 and 4.8m by 1911 (2.4). As population growth accelerated from the mid-eighteenth century, it was accompanied by regional changes in distribution which reflected differential economic growth and variations in social and population behaviour (Deane and Cole, 1962). While the Scottish Highlands remained sparsely populated and vulnerable to food crises, growing towns and industries in central Scotland absorbed much of the country's growth in the late eighteenth century.

Early British censuses provide numbers, density and change in population every ten years for parishes, counties and urban areas. There was some undercounting, especially in 1801. The density maps of England and Wales use the 624 registration districts created when civil registration was established in 1837. Civil registration came to Scotland in 1855, but the equivalent districts, 900 of them, were too complex, so counties have been used in these maps. In England in 1801 (2.1), areas of high density population were isolated and very small. They were provincial capitals and commercial centres. Especially prominent were London and the industrial west Midlands, south Lancashire and west Yorkshire. Whilst the countryside between East Anglia and the West Country was well populated, early proto-industrial areas in northern and midland England were growing faster (2.5).

By 1851 population density had increased throughout Britain (2.2). Although the uplands lagged, a growing, still labour-intensive agriculture supported by a substantial range of craft industry pushed rural population densities to their peak. The distribution was dominated, however, by towns and industry, especially coalfields, major ports and commercial centres. London's county with a population of 2.7m equalled that of the eleven biggest provincial cities put together, though expanding high density areas point to emerging conurbations around Birmingham, Manchester, Liverpool, Leeds and Glasgow.

Between 1851 and 1911 rural populations declined throughout Britain. In England and Wales rural numbers fell by 0.6m, from 46 to 21 per cent of the total. Densities fell severely in the marginal areas, especially the Highlands and Southern Uplands of Scotland and the uplands of Wales and northern England. In contrast, on the coalfields, especially northeast England, the east Midlands and south Wales, and in the rural/suburban periphery of large towns and conurbations, densities increased rapidly. A century of growth and redistribution had created a very different map from that of 1801 (2.3). On the coalfields and around London there were large densely populated areas. Suburban growth and conurbation were now appropriate descriptions of what was happening. The growth of older industrial areas in northern England, the Black Country and on Clydeside began to slacken, whilst that of Greater London, the southeast and the south and east Midlands quickened as population began the twentieth-century drift to the new industries of southern Britain (2.6).

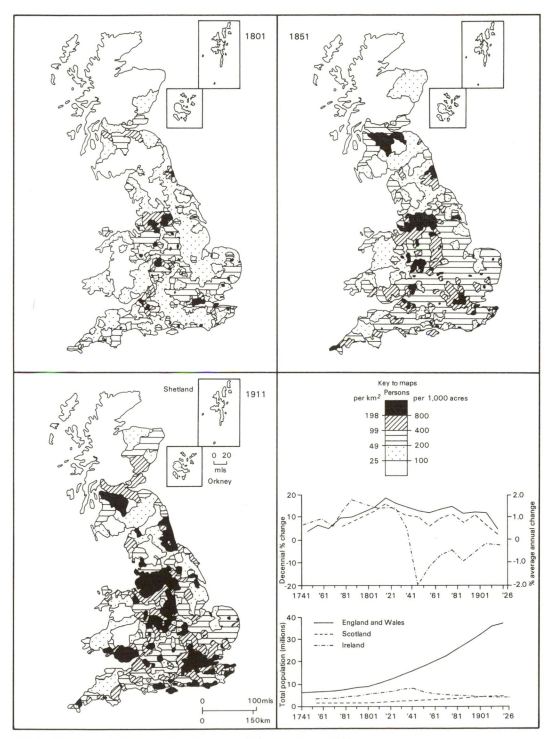

2.1–4 Population density, 1801–1911, and change, 1741–1926

The dominant nineteenth century trends were urbanization and growth, and after mid-century increasingly extensive rural depopulation (2.5–6). The late eighteenth- and early nineteenth-century parish register data suggest substantial regional variations in fertility and mortality (Lawton, 1977). Natural increase (births minus deaths) was high in many rural areas but greatest in the industrializing regions, despite high mortality in large towns where deaths often exceeded births until the early nineteenth century. By the mid-nineteenth century systematic civil registration from 1837 and 1855 and the fuller censuses from 1841 permit a more accurate analysis of the components of population change. Under-registration was still a problem. For England it was 9.4 per cent in 1840–5 and 2.4 per cent in 1860–4 (Glass, 1973).

Between 1801 and 1851 London's population growth (960,000 to 2.7m) owed much to migration, since in central areas of the city deaths generally exceeded births. The high growth areas were the coalfields, the industrial west Midlands, south Lancashire and west Yorkshire, high general and epidemic mortality being offset by high fertility (2.5). Increased demand for agricultural labour and continuing work for craftsmen and their families produced a peak of population increase in most rural parishes between 1811 and 1821, parts of the Scottish Highlands excepted. But in remote areas – upland Wales, the northern Pennines and northern Scotland – early migrational losses led to decline from the 1830s. In parts of Scotland, eastern and southern England and coastal areas of Wales, rural increases still matched the national average.

As agricultural employment declined and industry was progressively absorbed into larger urban production units, rural populations fell between 1851 and 1911 (Lawton, 1983). Many parishes had fewer people in 1911 than in 1801 (2.6). Clearances further reduced natural growth in northern Scotland and, to a lesser extent, the Borders, though natural increase in northeast Scotland often matched that in central Scotland where high fertility in industrial areas was offset by high mortality. In rural England, except for upland areas, natural increase was still near national levels, until the prolonged outflow of young people led to reduced natural increase in the 1890s. High fertility was offset by high mortality in many large towns and heavy industrial districts such as the Black Country and the Potteries. In the coalfields the balance of high fertility and mortality produced high rates of natural increase. In the suburbs, such as those around London, high birth rates and low mortality produced the high natural increase.

Much of the contrast in growth was due to differential net migration. In Map 2.8 the migration component was measured as the difference between natural increase and the actual change recorded between censuses. All types of rural areas lost heavily, though net outmigration was greatest in the uplands. Areas of migrational gain were few, especially in Scotland where only Edinburghshire, Lanark and Dumbarton gained consistently. English cities attracted large numbers as levels of urbanization increased (Section 22). Indeed increases in urban and industrial population absorbed 91 per cent of the country's 21.4m natural growth in the period 1841–1911. But areas of persistent migrational gain were concentrated in the suburbs of London and the larger cities. Many older northern and coalfield industrial areas had relatively modest increases, and by the late nineteenth century were losing population to newer growth areas in the Midlands and southeast.

Underlying all sex ratios (2.9–12) was the male surplus at birth which was reduced and reversed by differential infant mortality and the hazardous male occupations in heavy

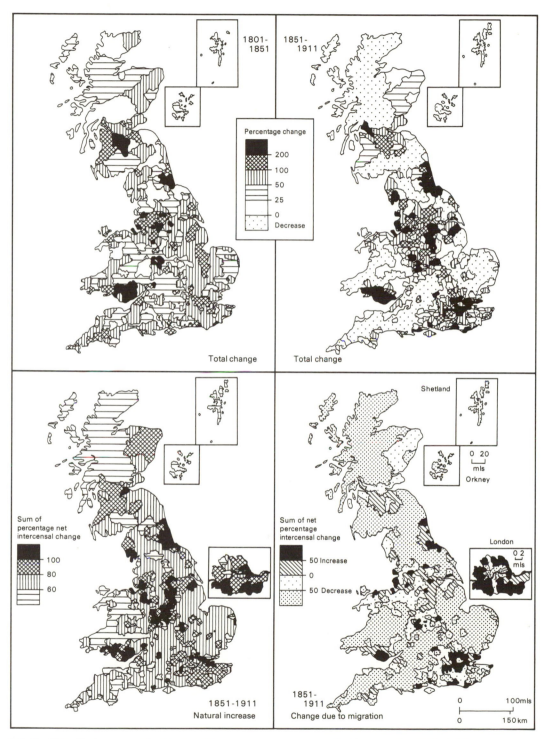

2.5–8   Population growth, 1801–1911; components of population growth, 1851–1911

industrial areas. Sex ratios are conventionally calculated as males per 100 females. Reporting of gender in British censuses was considered reliable. The eighty-eight registration counties are the most practical unit for mapping these data. Since the range of values changes over time, all maps of structure and vital trends are mapped by quintiles. This divides the occurrences into five equal groups and eases comparison of patterns at different dates.

Migration left a male bias in the countryside, as women had overall a greater propensity to move into towns for work and marriage (Saville, 1957). Sex structure reflected employment opportunities, with a male bias in ports, mining and metal-working areas and female dominance in textile districts and middle-class residential areas employing domestic servants. The greater propensity to migrate in the late teens and early twenties meant that young adults dominated areas of active immigration, and older groups areas of persistent loss. Life-stage residential migration creates sharp differences in both the sex and age structure between different types of town and between different parts of large towns, as female dominance of retirement and residential areas indicates.

In 1801 (2.9) the most striking general feature was the low M:F ratio (92 in England and Wales; 85 in Scotland) reflecting the exclusion in the census of men under arms. Comparison with later census data is thus precluded. Most rural Scottish counties had very low ratios, as did the two northern counties and much of southwest England and southwest Wales. In contrast most of rural eastern England and the Welsh borders were above average. High ratios in the southeast suggest substantial female migration to London. Most industrial counties had above-average ratios, especially the heavy industrial west Midlands and Monmouth, though male dominance was less marked in Lancashire where there was already employment for women in textiles. In 1851 (2.10) lower ratios persisted in Scotland, reflecting high male emigration, but areas of substantial male employment (e.g. ports, the coalfields and Clydeside) were at or above average. Two types of male-surplus area (above-average ratios) persisted in England: first, heavy industrial counties like Staffordshire, Derbyshire, Durham and the West Riding; secondly, rural areas like the Welsh borders and a belt from Lincolnshire to Kent. The major cities, especially London, had low ratios though these tend to average out in surrounding counties. The below-average ratios of southwest England and west Wales reflect migrant males. In 1911, differences between England and Wales and Scotland were less (2.11) because of differential changes in the ratio (2.12). This provides a basis for analysis of Britain as a whole in a period when the exodus from the countryside to the town and shifts in location of industrial activity were reshaping the economic map. The main features are the growing levels of female dominance (low ratios) in major city-regions, especially London, Bristol, Liverpool, Manchester, the textiles towns, Edinburgh and the Scottish borders. Male dominance continued in some rural counties especially of the western Highlands, eastern England and mid-Wales, but was most marked in the coalfields, some ports and the heavy industrial areas.

Britain's youthful population structure in the early nineteenth century reflected relatively high fertility and mortality. Pre-census sources are of little value for regional age data. England's proportion of young dependants increased after 1750 with rises in birth rate, but was falling again by 1850 (Wrigley and Schofield, 1981). Scottish data suggest a youthful population. The Old Statistical Account gave 25 per cent under 10 years old and 36 per cent under 20, whilst Webster's shaky arithmetic gave similar

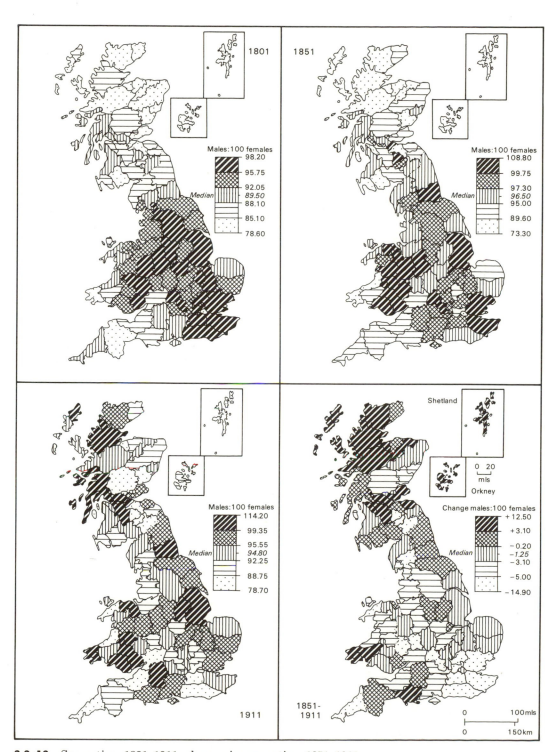

2.9–12   Sex ratios, 1801–1911; change in sex ratios, 1851–1911

proportions in 1755 (Flinn *et al.*, 1977). Only with fertility decline and improved urban mortality in the late nineteenth century did the proportion of the population under 15 years fall and that over 45 rise.

The 1851 census had the first usable age data. Inaccuracies certainly occurred – for example understatement of infants, overstatement of younger female ages and rounding down – but age data were reliable by the late nineteenth century. No single index exactly portrays comparative age structure. Population pyramids are a simple device for individual areas but are difficult to map at small scales. Here, the proportions of four groups – children under 15, younger (15–44) and older (45–64) working ages and elderly (over 65) – in registration counties are mapped by quintiles.

In 1851, Britain's population was still youthful, 36 per cent being under 15 (2.13–16). The least youthful areas had experienced long-standing outmigration and low birth rates; northern Scotland, the Highlands, parts of southern England and mid-Wales. But in large cities, especially London, high mobility of single persons was associated with fewer children. High fertility rural areas, such as the east Midlands, southeast England and west Wales, had above-average values. High proportions of working-age population (15–44), reflecting substantial inmigration, characterized the industrial areas and London (2.14). Older groups were still small nationally – 14.6 per cent aged 45–64 and 5.5 per cent over 65 – reflecting rapid population growth. Higher proportions were mainly in areas of marked rural depopulation such as northern and highland Scotland, west Wales and counties of southern England (2.15–16). In 1911, (2.17–20) rising life expectancy and the onset of fertility decline was reflected in smaller percentages of under-15s (30 per cent) and larger proportions of adults from the 1890s. Losses of young adults from rural areas, inner cities and some older industrial regions, and movement into outer suburbs and new industrial areas, were reflected in more clearly defined regional patterns. The areas of most youthful population, central Scotland, industrial northern England, the west Midlands, south Wales, London and the southeast, were generally also those with the lowest proportions of the elderly, though Sussex and Devon were above average in both groups. The highest younger working-age populations were still in urban/industrial counties (2.18): central Scotland, Aberdeen and Forfar; northeast England, the Midlands, south Wales and Greater London. But there were relatively more middle-aged in residential areas like the outer southeast.

This complex picture suggests a threefold grouping: large cities (working-age dominance) and industrial areas (more children and younger adults); rural areas with ageing populations, some with above-average 0–14 age-groups; finally, populations whose behaviour was close to the national average, including parts of the home counties, south Midlands and west of England.

Indices of fertility and age-specific death rates point to growth from the mid-eighteenth century (2.21). Death rates fell between 1750 and 1820 from 26 to 22 per 1000 in England and in Scotland from rates possibly as high as 38 to 20 per 1000 in 1855 (Wrigley and Schofield, 1981). There were substantial improvements in infant mortality. Rising fertility reflected a fall in marriage age, higher nuptiality and illegitimacy. English crude birth rates increased, 1770–1820, from 35 to 41 per 1000, falling again to 35/36 between 1830 and 1870. By 1914, fertility decline from the mid-1870s had reduced English birth rates to 24 and those in Scotland to 26. Average family size fell from 6.2 children in marriages of the 1860s to 3.3 for the marriages of the 1900s. The spread of family limitation did not eliminate regional fertility differentials. (2.27).

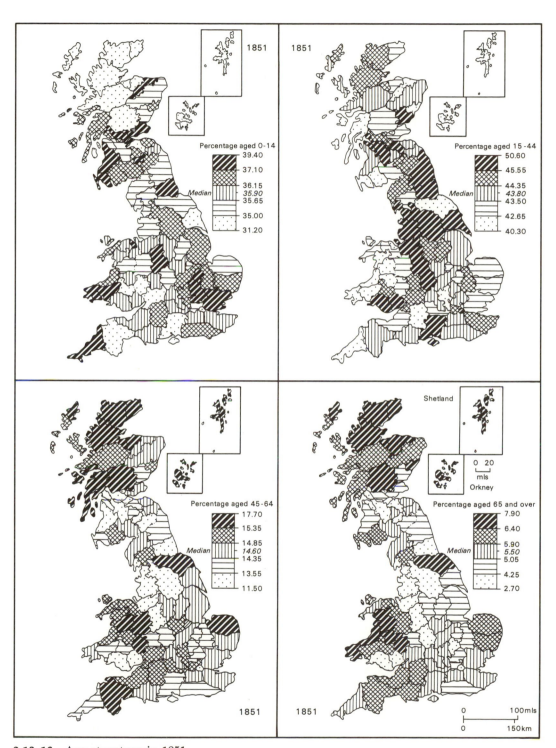

2.13–16 Age structure in 1851

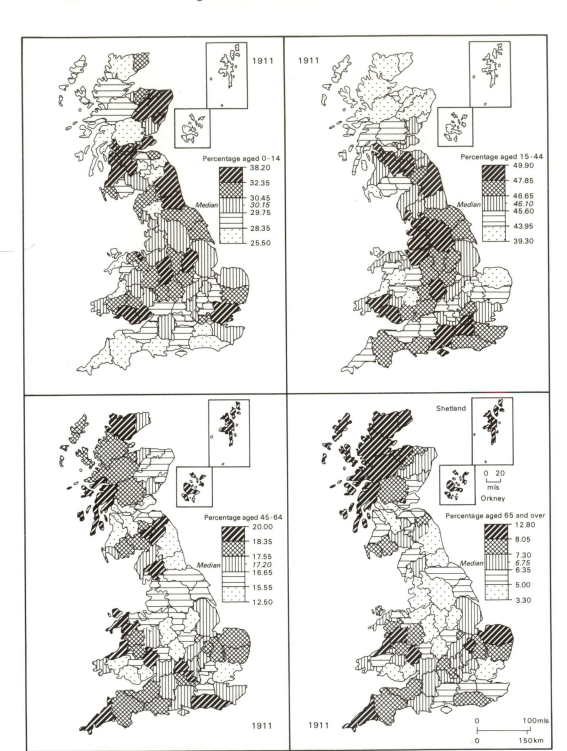

2.17–20 Age structure in 1911

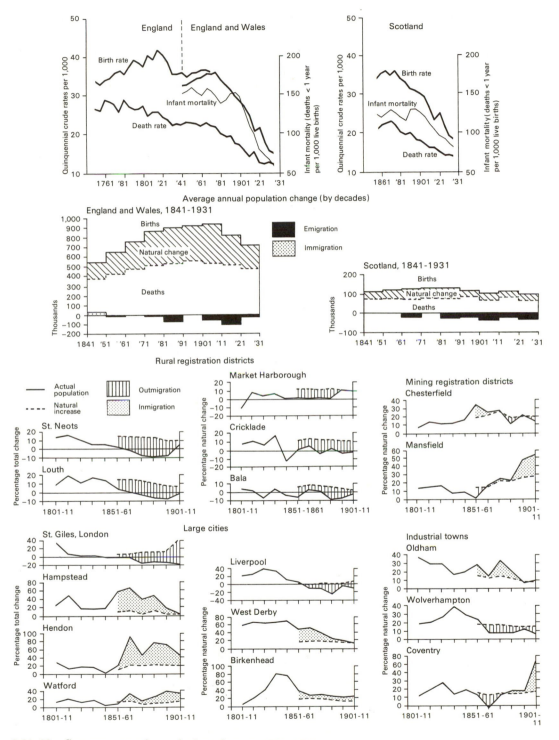

2.21–23   Components of population change, 1801–1931

Crude death rates changed little between 1820 and 1870. Unhealthy urban and industrial areas were prey to epidemics (cholera; typhus) and the endemic diseases of poverty and overcrowding – tuberculosis, intestinal and pneumonic infections – which offset rising life expectancy in rural areas, keeping national rates to 21 to 23 per 1000. From the 1870s mortality declined among children and young adults. From the 1890s, older age-groups were affected as better public health, nutrition and hygiene lengthened life. Infant mortality remained high until the 1890s, then fell sharply from 150–60 to around 100 in 1914 in England and Wales and from 117–38 to about 110 in Scotland.

The overall result of this was substantial natural increase with fluctuations shaped by migration (2.22). Except during the Irish famine migration of the 1840s, Britain was a net exporter of people. Scotland's net loss was 3.5 million over two centuries after 1750. England and Wales lost 1.4m, though England was a net gainer from movements within the British Isles, especially from Ireland: Scotland gained from Ireland but lost to England.

At local level the detailed impact of economic change was evident (2.23). Rural England and Wales had net migrational losses of 4.5m, 1841–1911. Registration districts on both clay (St Neots) and light soils (Louth) lost heavily, especially from the 1870s depression. Fatstock (Market Harborough) and dairying (Cricklade) areas, though less vulnerable to overseas competition before 1900, also lost population. Hill farming (Bala) declined early and substantially (Lawton, 1968). Migration contributed a quarter of the gains in London and the eight largest towns between 1841 and 1911. Inner city districts like St Giles in London experienced high mortality and low natural growth and, from the 1840s, net outmigration in successive phases of suburbanization with migrational gains in outer districts like Hampstead in London and West Derby in Liverpool (Lawton, 1983). Older industrial districts lost some 9 per cent of their natural increase, 1841–1911, especially after 1870 (e.g. Wolverhampton). Textile districts augmented relatively low natural increases, gaining around 5 per cent of their growth from migration. They lost in the 1860s depression and like Oldham saw longer-term decline in the 1890s. Coal-mining areas, like Chesterfield and Mansfield, had a high but volatile natural and migrational growth. Coventry's population responded rapidly to new late nineteenth-century industries (sewing machines and bicycles) after the earlier slump of ribbon-weaving. Inmigration (mainly young adults), generally promoted natural growth but left depressed growth in areas of prolonged out-movement.

Before civil registration, spatial analysis of birth and death rates is very incomplete. Imperfect evidence for Scotland, England and Wales suggests wide regional variations in birth and death rates. The Highlands and the west of Scotland were high in both cases (Flinn et al., 1977). In England, London's death rates were high as were those of the northwest and Midlands. Birth rates were above average in the proto-industrial areas of the southwest and east Midlands, as they were in the industrial areas of the west Midlands and northern England.

Post-civil registration, average decadal crude birth and death rates per 1000 of total population have been mapped at county level, again by quintiles (2.24–27). In the 1850s, the dominant feature was the high birth rates in urban/industrial counties, especially the heavy industrial areas of central Scotland, northeast England and south Wales. Notice that the largest cities, especially London, do not have the highest rates. Some largely rural areas in northeast Scotland, the eastern Borders and the south Midlands of

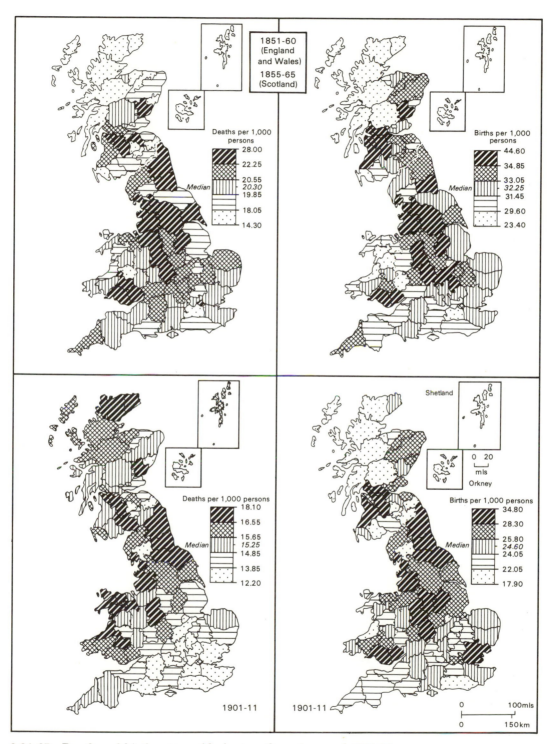

2.24–27   Death and birth rates, mid-nineteenth century and 1901–11

England had relatively high birth levels. Indeed most of eastern Britain was around or above average. In contrast, rural areas in southern England, Wales and, especially, northern Scotland and the Highlands had low birth rates reflecting outmigration. Detailed studies (by registration district) using more refined methods of analysis suggest a fertility gradient between rural and urban/industrial areas and from north and west to south and east in England, but with complex local variations reflecting socio-economic contrasts in family size (Woods, 1982).

In the 1850s (2.24) low death rates in rural counties contrast with high rates in urban industrial counties. Slightly above-average figures for the south Midlands, the southwest and East Anglia perhaps reflect population structure. In Scotland there is a marked contrast between very low rural and high urban mortality. At local level throughout Britain an even sharper contrast may be observed between countryside and town and between low suburban and high city-centre mortality (2.37–38). In the 1900s, despite rapid reduction in mortality in towns in the late nineteenth century and falling infant mortality, a rural/urban gradient persisted. Crude rates masked this in some areas of ageing populations like north and mid Wales, northern Scotland and parts of the Borders. A clear south and east to west gradient reflects better rural environments and greater regional prosperity. The lowest death rates were in residential suburbs and towns with high living standards and better child-care facilities as in southeast England. Highest mortality was still experienced in the large cities and industrial areas of south Wales, the Midlands, northern England and central Scotland. However, longstanding depopulation contributed to ageing populations and relatively high crude death rates in much of Wales and northern Scotland. Conversely birth rates in such areas were low, as they tended also to be in many rural counties of southern England. Youthful populations and high birth rates dominated industrial areas, especially those with high rates and youthful age of marriage like south Wales, Yorkshire and the northeast coalfields. Standardized measures underline this feature, but suggest that in many middle-class suburbs fertility was falling.

A major cause of higher fertility in late eighteenth- and early nineteenth-century England was a fall in mean age of marriage of up to four years, together with increased illegitimacy ratios (Schofield, 1985). After 1830, a rather older mean age of marriage prevailed. Scottish data were meagre before civil registration, though marriage seems to have been younger in the 1780s–1800s than in 1810–55 in both the Highlands and Lowlands, especially in towns. English marriage rates were always higher than Scottish before the twentieth century though this is offset by higher rates of illegitimacy and 'irregular marriage' and co-habitation in Scotland.

Two measures of marriage rate are used for the 1850s and 1900s period (2.28–31): first, persons marrying in particular years per 1000 of the total population, and secondly the percentage of women 20 years of age and over who are or have been married. The much lower level of nuptiality and different marriage practices in Scotland makes comparison of the distribution of marriage throughout Britain difficult. For example marriage rates per 1000 were 17 in England and Wales in the 1850s and 15 in the 1900s, as against 13 and 14 respectively in Scotland. In 1851 substantially more Scottish women over 20 were never married, 36 per cent as against 28 per cent in England. Among older age-groups the percentage ever-married increased to over 85 at age 45. In Scotland, unlike England, a smaller proportion of women than men were married, perhaps a reflection of a lower M:F sex ratio, and the higher proportion of Scottish widows, 15 per

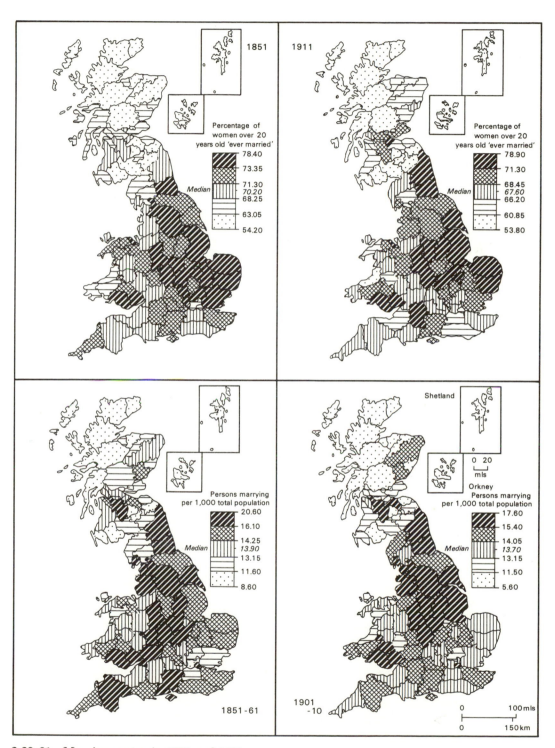

2.28–31   Marriage rates in 1851 and 1911

cent as against 13 per cent, reflects generally more adverse male mortality north of the border. The proportion of women ever-married was lowest in rural areas throughout Scotland and highest in industrial areas. While this was also the case in England, London like Edinburgh having low proportions, many eastern and midland counties and north Wales were above average.

Marriage rates in the 1850s (2.28) reflect these patterns. The highest proportions were focused on cities and industrial regions. Eastern and south coast counties of England were above average. In Scotland the urban/industrial areas were predominantly in the upper quintiles. Below-average rates around London and the low proportions in rural areas of Scotland, northern England and north Wales partly reflect mobility and its influence on sex and age structures. Large movements of single women into cities produced lower sex ratios and proportions of ever-married women, especially in Scotland and 'retirement' counties, but the figures were relatively high in rural areas of eastern England, the southeast Midlands and Wiltshire, often associated with balanced M:F ratios. In areas of female surplus, including central areas and middle-class city suburbs, ever-married percentages were often relatively low. Marriage rates in the 1900s were heavily skewed, reflecting low figures in rural Scotland and parts of Wales due to prolonged out-movement of young people. Most English rural areas were average or above, reflecting balanced sex ratios. But the highest rates were still in urban/industrial areas of central Scotland, south Wales and midland and northern England, reflecting youthful populations.

Pre-census evidence from Poor Law records, apprenticeship registers, local listings and parish marriage registers, together with a comparison of natural increase and actual change (2.1–8), suggests substantial migration to cities. There was considerable long-distance migration not only to metropolitan London and larger provincial cities, but also by skilled workers to new industrial centres. Most mobility was short-range and often temporary, but with a complex wave-like motion it gravitated towards towns and industrial regions. Thus, in Scotland, seasonal migration often associated with harvesting and fishing led to permanent movement to the central Lowlands. An 1841 census question on the county of birth of all enumerated persons gave the first direct information on migration. The 1851 question required parish and county of birth of those born in Britain, and country of birth for the others, notably the Irish. These were tabulated by civil county, changing later to civil and administrative counties (Drake, 1972). The maps show for 1851 and 1911, by quintiles, those born outside the county of enumeration as a percentage of its total population (inmigration) (2.32 and 34) and those living outside their county of birth as a percentage of all natives (outmigration) (2.33 and 35). Birthplace statistics pose many problems as a measure of migration. First, outmigration is understated because those living outside their country of birth (Scots in England and vice versa) are not enumerated. Secondly, more short-range 'movers' appear for small-area counties than for large. Thirdly, in large concentrations of population straddling county boundaries, some local movement appears as 'migration'. Finally, birthplaces show 'lifetime migration' – those who move at any previous time – and not (like parish registers) actual migration over a specific time.

In 1851, longstanding short-range rural movement to towns and industrial areas was reflected in above-average proportions of outmigrants, particularly in the south and east Midlands and southern England (2.33). There were relatively high rates of outmigration from parts of northeast Scotland, the southwest Highlands and the

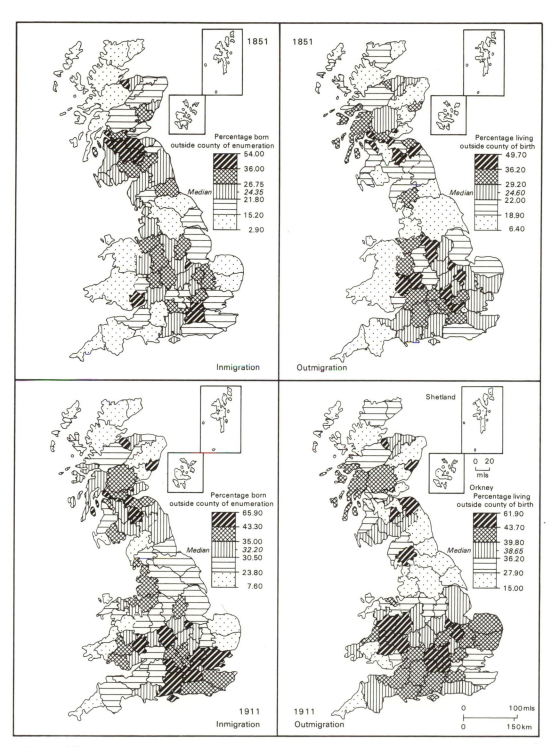

2.32–35   Migration in 1851 and 1911

Borders, but high inmigration to urban and industrial counties which were the beneficiaries of stage and stepwise movement from rural areas. The central belt showed the only net migration gain in Scotland (H. Jones, 1983) (2.32). Greater London took most of the migrants from the southeast and the south Midlands, while the west Midlands, though it gained from surrounding rural counties, also exported to northern England. Durham and south Wales mining areas recruited heavily from nearby areas.

Much unskilled industrial labour of the mid-nineteenth century was Irish. In 1851 a high proportion of the 0.52m Irish in England and Wales lived in cities, and 47 per cent of the 0.21m in Scotland were in the nine principal towns: one-quarter of Liverpool's population in 1851 was Irish-born (2.39). The 1911 evidence showed an increase in the total volume of lifetime migration. The inmigration median moved from 24 to 32 per cent and the outmigration median from 25 to 39 per cent. Largest in-movements focused on London and the southeast, the south Midlands and, less so, on the north and the north Midlands. South Wales continued to attract migrants, but northeast England produced more of its own labour force. The rural Midlands and southern England and mid-Wales had high outmigration as did the Borders and southwest Highlands of Scotland, though northern Scotland's levels were relatively low, perhaps reflecting migration from Scotland. Throughout Britain, city populations were moving into the suburbs, accounting for marked increases in migration, especially into the metropolitan counties. Migration is age- and sex-selective, and influenced population structure in both sender and receiver areas, emphasizing youthful structures in urban/industrial areas and ageing ones in rural areas, while life-cycle residential movements created distinctive demographic structures within the towns (Lawton, 1978a).

Differential vital and migrational experience created differing regional population patterns in nineteenth-century Britain, as the preceding maps demonstrate. Intraregional contrasts were often even greater than regional or rural/urban. Sub-regional analysis points to distinctive environmental, socio-economic and demographic relationships within communities. In pre-census periods these may be partly reconstituted from parochial and local records. From 1841 it is possible to link registration and published tabulations, and individual and household data are available from the census enumerators' books (Lawton, 1978b).

Liverpool's population grew dramatically in the late eighteenth century, from 34,407 in 1773 to 77,653 at the 1801 census, 375,955 in 1851 and 746,421 in 1911. Between 1801 and 1841 moderate natural increase, with high birth but increasing death rates, and high inmigration produced increases of over 20 per cent per decade.

From 1841, census and vital statistics indicate rapid growth in an arc of suburban parishes from Crosby in the northwest to Garston in the southeast and along the Wirral shore (2.36). The components of this change are shown for the 1850s and 1900s (2.37–38). Population declined with migrational losses in the city centre, where high mortality depressed natural growth. Suburbs grew rapidly with high natural increase in which populations were both youthful and healthier. There were substantial residential movements from the centre as well as large-scale inmigration (Lawton, 1979). Sharp contrasts in housing, social and environmental conditions distinguished inner and dockside areas, with relatively low birth rates and high death rates, and suburbs on both sides of the river, with high birth rates, relatively low mortality and very high migrational gain. Beyond the built-up area substantial natural growth in outer suburbs stands out from lower rates in rural parishes. From the 1870s slackening natural growth

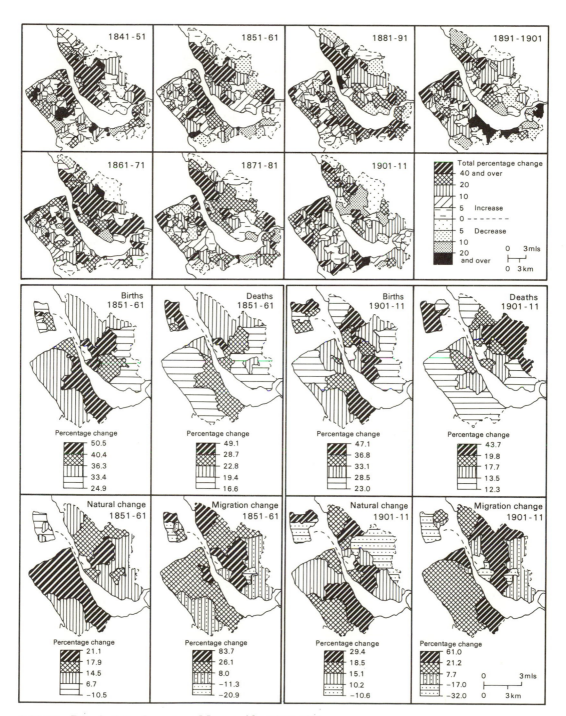

2.36–38 Population change on Merseyside, 1841–1911

and outmigration spread to central Birkenhead and Liverpool's inner suburbs. In contrast high levels of residential migration to newer suburbs are reflected in higher birth rates but increased mortality. Over the period as a whole some three cycles of change, with high increase, especially from migration, followed by slackening rates of growth and then decline, point to the outward movement and corresponding demographic and socio-economic changes within Merseyside.

By 1851, 174,000 (42 per cent) of Liverpool's population were born outside Lancashire and Cheshire (2.39–41). Most inmigration was short-range, but 22 per cent came from Ireland, 5.4 per cent from Wales and 3.7 per cent from Scotland. By 1901, the pattern was similar but with higher percentages (75 per cent) from Lancashire and Cheshire, mainly from within Merseyside. The numbers (45,673) and percentage (6.7) of Irish-born fell substantially, though not the proportions of Irish descent, while percentages of Welsh- and Scottish-born fell to 3.0 and 2.5 respectively. Decadal changes in migration may be estimated from comparison of lifetime migrants in successive censuses inflated by a survival ratio based on mortality rates for Liverpool. The pattern of new migration (over 60,000 in most decades) suggests declining rates of inmigration. Older links to Ireland, Wales, Scotland and northern England persisted, but relative growth in the movement from the southeast and the west Midlands points to inter-urban mobility, often of more skilled persons.

Migration streams differed in occupational, demographic and family structure. Some tended to concentrate in certain parts of the city: Scots merchants and professionals in high-class residential areas; Welsh builders and English artisans in bye-law terraces in expanding working-class suburbs; the poorest in the slums of inner and dockside areas (see also Map 22.31). Census enumerators' books for 1871 (2.42) show high concentrations of Irish in most north dockside areas which had over 30 per cent Irish household heads (up to 80 per cent of total population), with many lodgers and high levels of multiple occupancy: they were relatively highly segregated and mainly in unskilled manual work. In contrast there were few Irish in working-class residential areas inland from the docks and, especially, the middle-class sector southeast of the city centre.

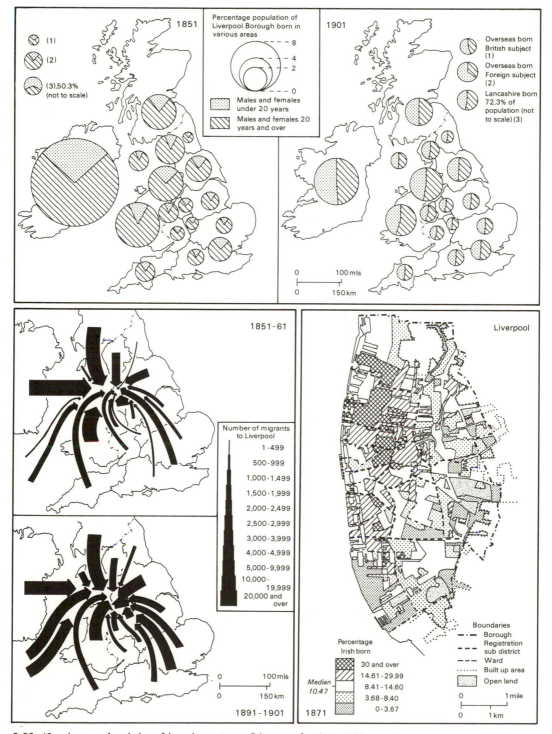

2.39–42   Age and origin of immigrants to Liverpool, 1851–1901

3 Regional structure and change

Clive Lee

The study of economic structure at regional level, prior to the very recent past, is limited by availability of information. It is not possible to construct a series of estimates of regional output, comparable to those which have been devised at national aggregate level. But the Census of Population, compiled at ten-year intervals since 1801, does allow the derivation of regional structures in terms of employment from 1841 onwards. The maps here show regional structure in employment at two benchmark dates, 1851 and 1911.

Employment has been divided into three components, agriculture (3.1–2), manufacturing including mining (3.3–4) and services (3.5–6). In fact, problems of definition and the format in which census data are published make the divisions, especially between manufacturing and services, less than clear-cut (Lee, 1979). But even at this low level of disaggregation, a number of important elements of structure and change during the Victorian period can be seen. The most obvious feature in this change was the fall in employment in agriculture both in absolute numbers and relative to the other two sectors. In Britain as a whole agriculture accounted for 22 per cent of employment in 1851, a small share by contemporary international comparisons and present-day Third World countries, but was less than 8 per cent by 1911. By the latter date it was only in the geographically peripheral areas of Britain that agriculture comprised a substantial share of regional employment. Manufacturing, by contrast, showed little change either in locational concentration or in share of national employment, increasing its share by some 4 per cent over this period. The pattern of service employment is perhaps less familiar, having been accorded less attention by most textbook writers. The share of this sector increased considerably, by some 10 per cent, during the sixty-year period shown here. Services thus absorbed most of the share of employment relinquished by agriculture.

Between 1851 and 1911 employment in Britain almost doubled. Rates of growth varied considerably between regions, but were greatest in the southeast and north. Yorkshire/Humberside, Wales and the northwest also enjoyed an increase greater than 100 per cent. East Anglia and the southwest fared worst with increases of 17 and 21 per cent respectively. This differential growth reflected the various regional structures. In Wales and the north coal-mining provided the main source of growth, while the slowly growing regions had a strong agricultural orientation. In the largest and fastest growing region, the southeast, the main stimulus to growth came from the service sector which increased threefold, although manufacturing employment more than doubled. Growth thus reflected and augmented the structural composition of most regions such that the scale of structural change was not great.

Apart from the decline in agriculture, a feature common to development throughout the western world, these maps show the consolidation of regional specialization during the Victorian years. While the comparative advantage of many regions lay in manufacturing or mining, others developed a strong specialization in services. Nowhere was this more evident than in the southeast where the economic structure of London spread to encompass the home counties. There thus emerged two distinct and quite different types of regional economic growth and structure in the nineteenth century (Lee, 1981). The same specializations can be perceived in embryonic form in the eighteenth century. To term this process an industrial revolution tells less than half the story.

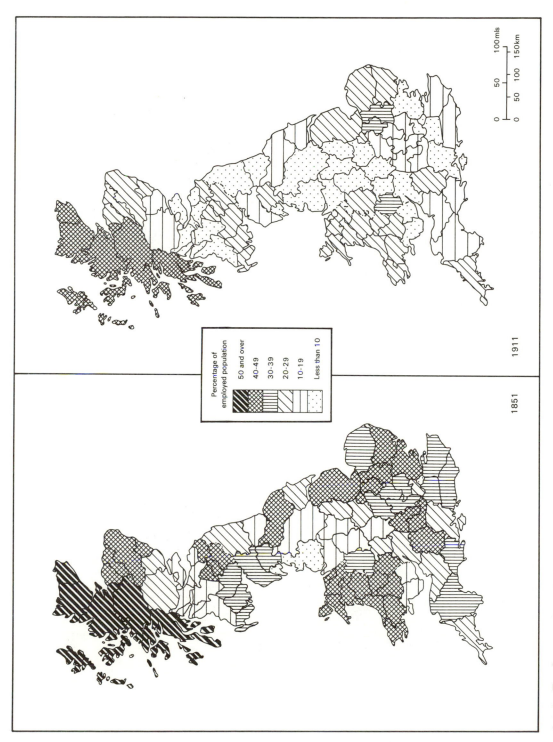

3.1–2 Employment in agriculture, 1851 and 1911

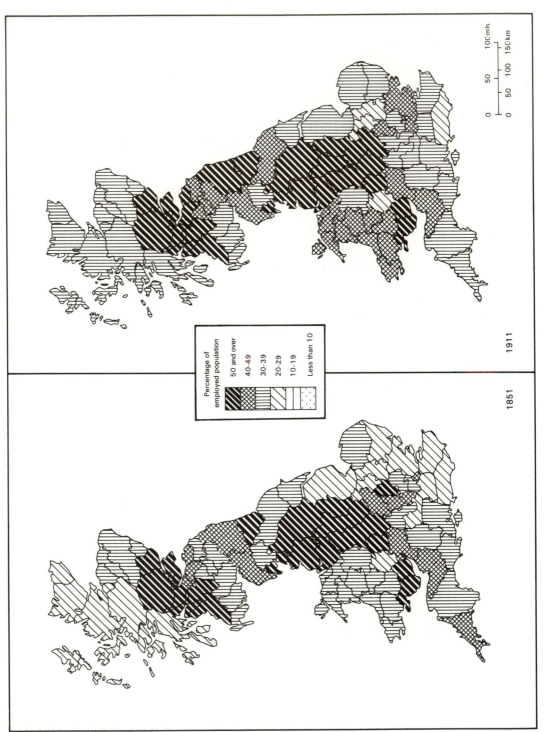

Percentage of employed population

- 50 and over
- 40-49
- 30-39
- 20-29
- 10-19
- Less than 10

1911

1851

3.3–4 Employment in manufacturing, 1851 and 1911

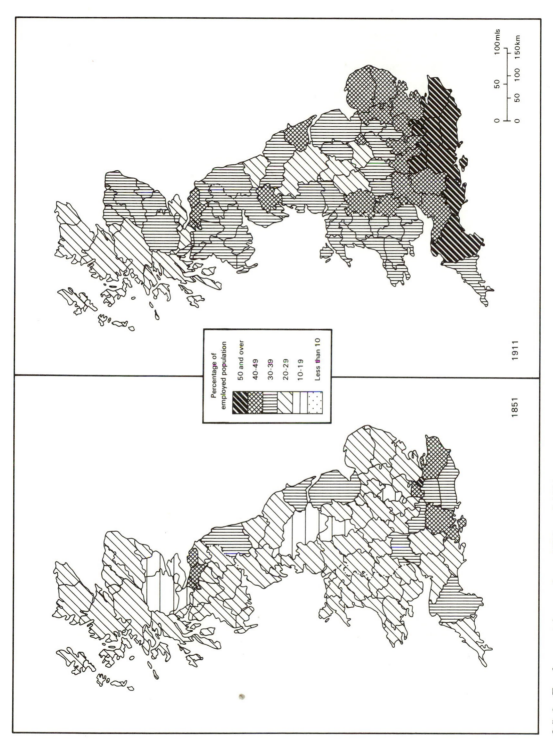

1911

1851

Percentage of employed population

50 and over

40-49

30-39

20-29

10-19

Less than 10

100 mls

50   100   150 km

50   100

0   50

0   0

3.5-6 Employment in services, 1851 and 1911

4 Agriculture
Mark Overton

As a consequence of industrialization the share of agriculture in the Gross National Product fell from some 45 per cent about 1750 to 6 per cent in 1911 and the proportion of the country's labour force employed in agriculture shrank from around 55 per cent in the mid-eighteenth century to 8 per cent in 1911 (3.1–2). Yet at the same time agricultural output experienced unprecedented growth. The population of 1850, three times that of 1750, was still largely fed by home production (Overton, 1987). Farmers responded in a number of ways to increasing demand. In the short term, during the crisis years of the Napoleonic Wars for example, output could be raised by increasing inputs or by extending the area cultivated. These strategies could not provide a long-term solution however, since comparatively little land was left to be reclaimed and increased inputs led to diminishing returns. Sustained increases in output could only be brought about by changes in farming technology so that more food could be produced from the same area of land. Technological change took many forms, from the rearrangement of existing systems of husbandry to the introduction of new crops and new systems of farming. Associated changes in the countryside ranged from the transformation of the physical landscape to the creation of new social relations between the groups involved in agricultural production. Thus two sets of processes transformed British agriculture from the sixteenth century onwards – changes in the technology of farming and changes in the institutional arrangements under which farming was carried out.

Changes in technology were directed towards increasing the productivity of land, through cropping innovations, and of labour, by the introduction of machinery. The major cropping innovations were two fodder crops, turnips and clover, which gradually became integrated into arable rotations. They increased livestock carrying capacity and therefore supplies of manure, the main fertilizer of arable land, which raised fertility and yields of grain. Turnips were first grown as animal fodder in the 1630s and clover cultivation dates from the early seventeenth century. Although data on the chronology and location of their introduction is patchy, we know for example that by the 1740s about half the farmers in Norfolk and Suffolk were growing turnips and about a quarter had clover on their farms (Overton, 1985), although it was not until after the mid-eighteenth century that these crops were having much effect on cereal yields. Turnips provided extra winter fodder but also acted as a cleaning crop if cultivated properly with regular hoeing. They were important in enabling some light lands, like the chalk downlands of southern England and parts of Norfolk, to be brought under the plough for the first time. The clover crop not only provided extra fodder, but as a nitrogen-fixing legume increased the supply of an essential nutrient for cereal crops. A third of the increase in arable productivity in northern Europe between 1750 and 1850 has been attributed to legumes such as clover (Chorley, 1981). Turnips and clover were often cultivated in a rotation where clover was undersown with barley, and turnips grown as a break between two grain crops. The ultimate expression of these principles was in the Norfolk four-course rotation of wheat, turnips, barley and clover, although the rotation was rarely practised in this pure form. Not all environments were appropriate for it, and even where soils and climate were suitable farmers usually wanted to grow other crops such as oats to feed their horses. The rotation also demanded an increase in livestock. Sheep fed off turnips in the fields during the winter months but the crop was often lifted and fed to bullocks housed indoors. Both turnips and clover enabled the frequency of bare fallows to be reduced since clover speeded up the process of nitrification and the cultivation of turnips cleaned the land, but some fallows were

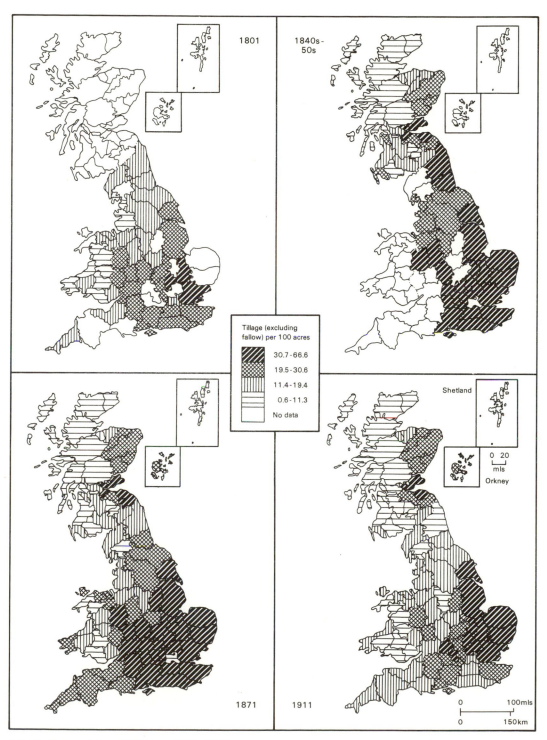

4.1–4   Percentage of tillage, 1801–1911

still necessary for eradicating difficult weeds until the introduction of herbicides in the present century.

Other new crops included potatoes, cultivated in increasing quantities from the mid-eighteenth century, swedes, which supplemented turnips, and grass substitutes including sainfoin, lucerne and ryegrass. From the late eighteenth century oil cake began to be used as fodder and some artificial fertilizers such as guano and coprolite were spread on the land. The import of these feedstuffs and fertilizers grew rapidly from the 1830s (Thompson, 1968).

The widespread introduction of machinery dates from the 1830s and 1840s. The 'new husbandry' made considerable demands for labour, particularly with the root crop, and as mixed farming intensified during the mid-nineteenth century into so-called 'high farming' (Jones, 1968), farmers looked for ways of reducing their labour costs. The main savings were made with the harvesting and threshing of grain. The first major change in harvesting was the switch from reaping grain with a sickle to cutting with a scythe, followed by the introduction of the reaper and the reaper-binder. The number of worker-days needed to harvest an acre of wheat fell from 4.8 with a sickle to 2.4 with a scythe and 0.5 with a reaper-binder (Collins, 1981). Threshing machines were introduced early in the nineteenth century to replace the laborious task of threshing with a flail (Macdonald, 1975). When wages fell, however, they were unpopular with both farmer and labourer and were often the object of attack during periods of unrest such as the Swing disturbances of 1830–1 (24.4). By the 1850s, however, they had become a permanent part of the rural scene, increasingly powered by steam. Five man-days were needed to thresh an acre of wheat with the flail, but only 0.8 using a steam-powered thresher. Other machines such as horse hoes, turnip cutters, winnowing machines, cake crushers and bean mills were also introduced (Walton, 1978) and constant improvements were made to existing implements such as ploughs. Between 1840 and 1900, output per worker in British agriculture probably increased by over 70 per cent. Yet changes in technology were not universal. With the exception of the use of steam power for threshing, horses provided the main power source in farming as they and oxen always had. The working practices of many livestock farmers of the west and north hardly changed at all (Jewell, 1975).

Most of these technological changes cannot be traced with the precision necessary to map them. Yearly series of national statistics of crops and livestock began in 1866 (Coppock, 1984) and it is only from that date that changing distributions can be monitored year by year. Crop yields were not collected until 1885, information on tenure until 1887 and questions on farm machinery were not asked until 1942. While it is not yet possible to draw a series of maps which cover the spectrum of agricultural innovation, it is possible to show some basic cropping statistics before 1866 which reflect some of the consequences of that innovation. In 1801 the government instituted an enquiry into the acreages under various crops in England and Wales, the data being collected on a parish basis by the clergy. Many of these parochial returns survive to enable us to reconstruct distribution patterns of seven arable crops in 1801, but unfortunately the returns have no information on grassland or animals. Similar information is also available on a parish basis in the tithe files which were compiled in the years after 1836 following the Tithe Commutation Act. The first consistent Scottish statistics are available for 1854 when the Highland and Agricultural Society administered a successful survey of Scottish farmers on behalf of the Board of Trade. These Scottish data are mapped with the

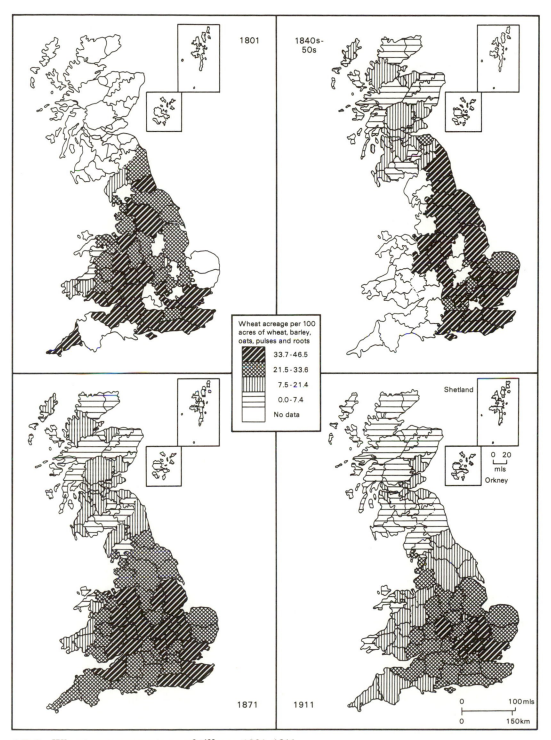

4.5–8 Wheat as a percentage of tillage, 1801–1911

tithe material and may therefore produce slightly misleading maps since the data for Scotland and England are separated by fifteen years or so. To the maps from these two 'cross-sections' are added two further ones for 1871 and 1911 from the agricultural statistics, giving four cross-sections in all. While in theory it is possible to follow William Marshall's advice of 1796 that 'in examining a country, like England, with a view to the existing state of its agriculture . . . the arbitrary lines of counties are to be wholly disregarded', it has regrettably been ignored because of the nature of available sources.

The intensity of arable cultivation over the century is indicated in 4.1–4. These maps share a common key which enables differences between maps to be distinguished as readily as the differences within each individual map. The class intervals, as with most of the other maps in Section 4, are determined by dividing the extant county data into quartiles; thus there should be an equal number of counties in each class interval over the four maps. So that all four maps may be directly comparable, 'tillage' is defined as cereals (wheat, rye, barley and oats), pulses (peas and beans) and roots (mostly turnips). The general features common to all four maps reflect variations in climate and to a lesser extent relief and soils. Comparison of 4.1–4 with 1.6–11 clearly indicates the greater suitability of arable farming to the drier and warmer areas of the country. While the physical environment can set broad limits to the range of crops that can be grown the changes over the century were a product of economic changes. Rising prices reflecting increased demand stimulated the expansion of the tillage area between 1801 and the 1840s, although the figures for 1801 may underestimate the true extent of tillage. Cheap grain imports after 1870 made cereal farming less profitable with the result (4.3 and 4.4) that the arable acreage fell. During this 'agricultural depression' many central and western counties of England turned over to grass but this was less easy in the drier east which continued to concentrate on grain farming. This process reflected a widening of the market resulting from reduced transport costs and the increased speed of information flows. There may already be signs of this process getting under way between the 1840s and 1871 with the reduction in the tillage area in Northumberland and Durham.

Maps 4.5–16 show the share of tillage accounted for by the three main grain crops: wheat, barley and oats. They show the proportions of tillage under the respective crops and do not necessarily reflect the quantity of the crop in a particular area. Thus oats formed a high proportion of a low density of tillage in northwest Scotland in 1911 (4.1 and 4.16). Wheat was the principal bread grain in England by 1800 with roughly twice the acreage of barley and slightly more than oats. Its widespread distribution in 1801 (4.5) is testimony to the general scarcity of the war years. Wheat prices had been rising steeply from the mid-eighteenth century but the 1790s saw a run of bad harvests and high prices. By 1871 there seems to have been a reduction in the proportion of tillage under wheat, giving some credence to the view that the repeal of the Corn Laws in 1846 did have some effect, however slight, on the home production of wheat (Fairlie, 1969). The impact of the depression after 1871 was most severe on wheat farmers and the crop became concentrated on the heavier lands in the southeast (4.7 and 4.8).

Like wheat and oats, barley can be grown under a wide variety of conditions although it prefers the light non-acid soils of the south and east. Its distribution reflects these environmental influences but also mirrors changes in the cultivation of wheat. Although the acreage of barley fell after 1871 (4.11 and 4.12) the fall was less than that

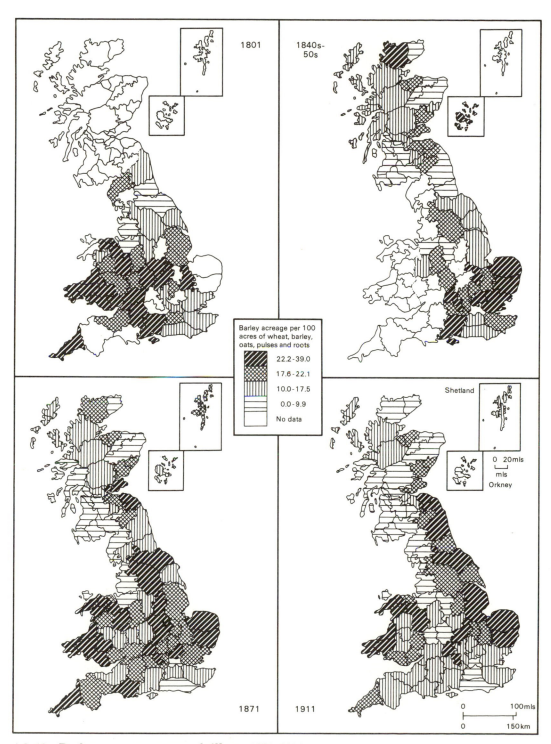

4.9–12 Barley as a percentage of tillage, 1801–1911

for wheat and the trend towards concentration was less marked. Maps 4.13–16 largely vindicate Dr Johnson's definition of oats, 'a grain, which in England is generally given to horses, but in Scotland supports the people'. It also formed a major part of the diet for many people in parts of northwest England and Wales. Oats are tolerant of a wide variety of soil and climatic conditions and the distribution of the crop is a result of its competition with wheat and barley. Oats were widely grown as a fodder crop and as the profitability of wheat, and to a lesser extent barley, fell after 1871 both the absolute acreage of oats and its proportion of tillage grew (4.15–16).

By the mid-nineteenth century these cereal crops were grown in many places as part of a rotation embodying the principles of the Norfolk four-course. Maps 4.17–20 suggest where that rotation was most dominant. Root crops contributed more than 25 per cent of the tillage area (which excludes sown grasses like clover) in the arable counties of Norfolk and Hampshire by the 1840s and by 1854 in the arable counties of eastern Scotland. Although they continued to gain ground as the century progressed, parts of arable England, most notably the heavy soil areas Essex, Huntingdon and Bedfordshire, never saw them contributing more than 13 per cent of the tillage acreage. Roots were also important as a fodder crop in areas far removed from Norfolk arable rotations. Thus they formed a high proportion of tillage in the livestock counties of the southwest and northwest of England and southwest Scotland. If root crops can be taken as an indicator of agricultural progress then Maps 4.17–20 indicate that they had much ground still to conquer in 1801. The acreage of root crops, in the twenty English counties with complete data for all four periods, probably doubled over the century and rose from some 11 per cent of tillage in 1801 to 17 per cent in the 1840s, 21 per cent in 1871 and 22 per cent in 1911.

The dangers of reducing the industrial revolution to a 'wave of gadgets' apply equally to characterizing agricultural change as a series of innovations which marched inexorably across the countryside, instantly transforming an ancient rural economy into a modern one. Technological change cannot be understood in isolation from the people who were responsible for introducing it. More particularly we need to sketch out the changing relationships between those involved in farming and the institutions which constrained their actions. We must also take care to avoid the simple-minded association of particular innovations with particular progressive (and often aristocratic) heroes who triumphed over a reactionary mass of country bumpkins.

Farmers in the mid-eighteenth century were limited in their freedom of action in several ways. Those in much of midland England had their arable lands scattered over the village in long narrow strips so that their 'farm' consisted not of a series of fields surrounded by a ring fence but an assortment of land parcels distributed across a series of large 'open-fields'. The 'open-fields' were so called because each contained a large number of strips, the basic unit of cultivation, with no physical boundaries around them. Village rules and regulations governed the crops grown, how many animals farmers could keep, and when such basic operations as ploughing and harvesting should take place. Each village could set its own rules and enforce them through local courts. In many villages the concept of private property as we know it today did not exist, in that the exclusive right of ownership did not give exclusive rights of use (Dahlman, 1980). Thus a landlord might own a tract of rough scrubland or pasture, often called 'waste', but the inhabitants of the village would have the right to graze their animals on it or to gather firewood. These common rights could also exist over the 'open-fields',

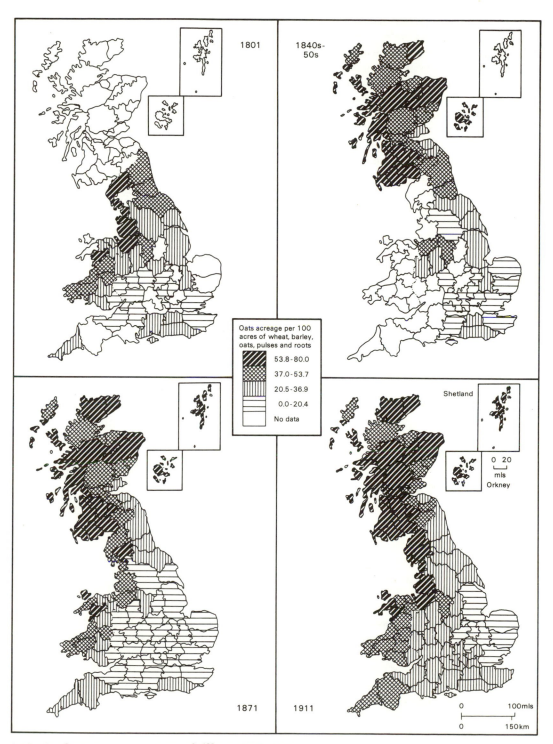

4.13–16 Oats as a percentage of tillage, 1801–1911

most typically when the owners of animals had the right to graze their beasts on the stubble of all the strips in a field after the harvest. When open-fields were subject to these common rights they were often called 'common-fields', though not all open-fields were subject to the same common rights and common right existed over many other types of land. In Scotland field systems were similar to the midland England model described here, although strips were known as rigs and land was cultivated in 'infields' often surrounded by the 'outfield' of land cultivated less intensively (Dodgshon, 1981).

Many of these features of field systems were changed by the process of enclosure (5.1–2). The actual word 'enclosure' can mean many things including the physical rearrangement of land and the creation of much smaller fields surrounded by a permanent boundary; the removal of common rights and the establishment of private property rights; and an effective end to co-operative husbandry so that each farmer was free to determine the nature and management of his farm enterprise. Much enclosure took place 'by agreement', whereby villagers could agree amongst themselves to enclose part of their lands, or acquiesced to their landlord doing so, but from the 1750s onwards enclosure of open-fields and of 'waste' areas took place in England and Wales through the mechanism of a private Act of Parliament. Map 4.21 reflects those areas of England which were still under open-fields in 1750 and Map 4.22 those parts of England and Wales where there were still extensive tracts of common land. It must not be assumed that areas largely untouched by Parliamentary enclosure never experienced open- or common-fields. In some places enclosure preceded the era of the Parliamentary Act. Parliamentary enclosure continued for about 100 years and by 1850 virtually all arable land was enclosed; open-fields had disappeared, private property rights were established and the management of farms was left to individual farmers or their landlords (Snell, 1985). Before enclosure some inhabitants of a village would have common rights over certain areas even though they might not own any land. Small allotments of land were often made to them in lieu of common rights so that the distribution of allotments on Map 4.37 is closely related to the distribution of open-field enclosure. A similar process of enclosure took place in Scotland during the same period but a series of Parliamentary Acts in the 1690s had paved the way for enclosure to take place without a specific Act for each village.

By the 1850s, or even earlier, the uniquely British pattern of social relations had emerged in the countryside. Most of the land of England was owned by landlords who leased land to tenant farmers who in turn hired farm labourers. Hardly any subsistence farmers remained except perhaps in western Scotland and the proportion of farmers who were owner-occupiers was probably below 15 per cent. This three-tiered division of capitalist farming: landlords, capitalist tenant farmers and agricultural proletariat, replaced a multiplicity of previous relationships. This required the concentration of land ownership into fewer hands, the replacement of varied tenure types and conditions by leases for a term of years, and the availability of a body of wage labour. Marx linked these changes to the period of Parliamentary enclosure although most historians would now locate them in the century after 1650 (Tribe, 1981).

Map 4.23 charts the distribution of wage labour in agriculture in relation to family farms, or rather farms not employing labour. The 1831 census provided the first information on the geography of this key feature of agrarian capitalism. The map strikingly demonstrates the dominance of wage labour in both southeast England and southeast Scotland. It is of particular interest that Maps 4.21 and 4.23 are sufficiently

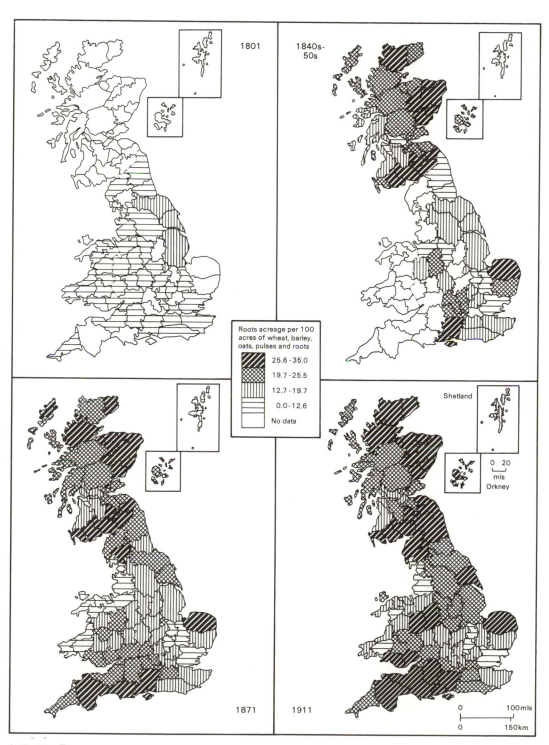

Roots acreage per 100
acres of wheat, barley,
oats, pulses and roots

25.6 - 35.0
19.7 - 25.5
12.7 - 19.7
0.0 - 12.6
No data

1801

1840s-
50s

1871

1911

Shetland

Orkney

0   20
mls

0   100 mls

0   150 km

4.17–20   Roots as a percentage of tillage, 1801–1911

dissimilar to reject the notion that Parliamentary enclosure alone was responsible for the establishment of an agricultural proletariat. Map 4.24 adds another dimension to the condition of those working on the land. Farm servants lived in with the farmer and his family on a yearly contract which included board and lodging. Although the proportion of labour hired as servants was falling rapidly by 1831 the map does indicate that in parts of Wales, northwest England and Scotland over half those employed in agriculture were living-in servants rather than labourers hired by the day or week (Kussmaul, 1981).

The incidence of agricultural labourers in part reflects the distribution of arable farming (4.3) which has very different labour requirements to livestock husbandry, but it is also a product of variations in farm size. Statistics on the sizes of farms were first collected on a national scale as part of the 1851 census but these data are not very accurate. There is no regular consistent information on farm size in the agricultural statistics although sporadic data is available from 1870. Maps 4.25–28 show the distribution of farms in four size groupings using the statistics taken in 1895 which have been chosen because they are the most comprehensive. The relationship between the density of agricultural labourers and large farms (4.23 and 28) is clear and the maps also indicate quite subtle differences in farm sizes which reflect particular histories of landownership. While there is considerable variation among the English counties it is noticeable that northwest Scotland (4.25), Wales (4.26) and central southern Scotland (4.27) display a greater unity of pattern.

These changes in social relations are naturally related to the production changes discussed earlier. Some argue that the innovation of new crops was dependent upon changes in property rights, although it could also be argued that technological change provided the impetus for changes in the relationships of those involved in the production process. In any event both sets of processes can be regarded as opposite sides of the same coin: the short-run stimulus was greater profit and the long-term consequences were increases in agricultural output and a transformation of social relationships in the countryside.

The opportunities for increased profits were more limited after 1870; indeed many farmers faced the prospect of falling profits. The vigour of British agriculture is supposed to have waned in the face of cheap imports of wheat from the 1870s. Wheat prices fell by 50 per cent between the early 70s and the mid-90s, and meat prices from the 1890s. Historians have disputed the meaning and significance of this 'depression'. The prices of other grains did not plummet as far as those for wheat, and while some meat prices fell, imported meat was generally of a lower quality than that produced at home. Maps 4.29–40 supplement the tillage maps (4.1–20) and confirm the differing regional experiences of 'depression' suggested in standard accounts (Perry, 1973).

Maps 4.29 and 4.30 demonstrate a general move away from tillage leaving only the eastern counties with less grass, both permanent and temporary, than tillage by 1911. The increase in the number of livestock is shown in Maps 4.31 and 4.32. These are density maps so that in conjunction with the previous two maps they can show the distinction between those areas with most of their cultivable land under grass but a comparatively low density of animals like Argyll, and the intensive mixed farming areas like Lincolnshire with both a high proportion of tillage and a high density of livestock. The dominance of sheep in Map 4.35 demonstrates their suitability for most farming environments; from the fattening area of intensive mixed farming systems in the east to

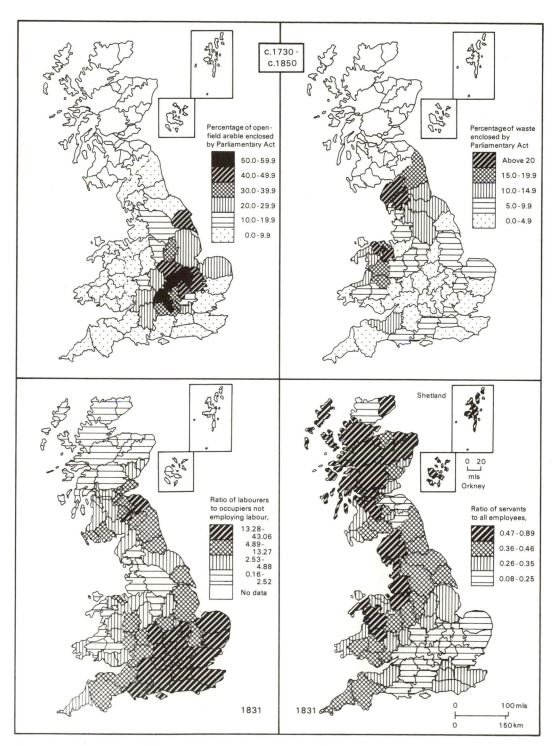

4.21–24   Enclosure, *c.* 1780–*c.* 1850; labour supplies, 1831

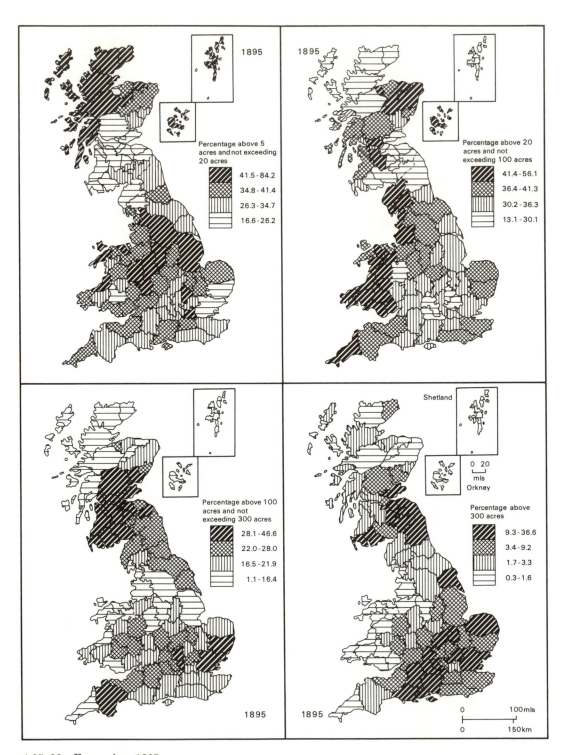

4.25–28   Farm size, 1895

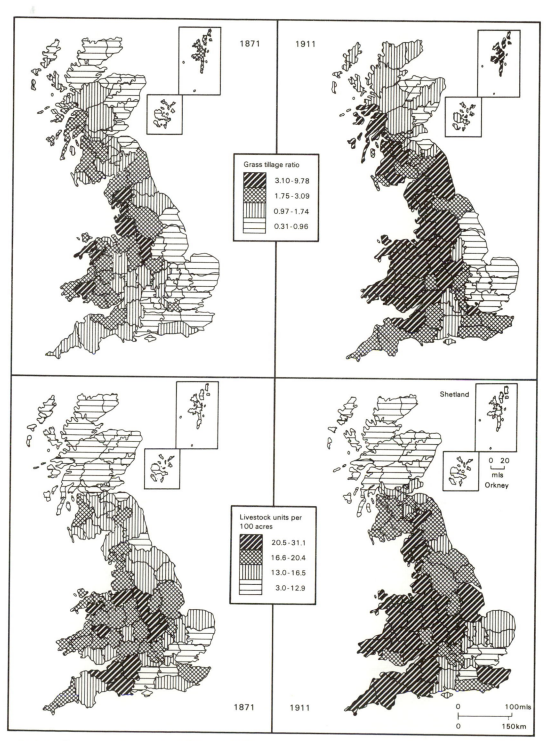

4.29–32   Grass and livestock, 1871 and 1911

the rearing areas of the mountains of the north and west. Although the numbers of sheep fell between 1871 and 1911, as English producers reacted to cheap imports of mutton and wool, sheep remained the most dominant animal in British farming. In contrast to sheep the relative importance of dairy cows was growing. Maps 4.33 and 4.34 show that the established dairy areas of northwest England and southwest Scotland were joined by an area of southern England. Liquid milk was immune from foreign competition and the railway enabled London to be supplied with fresh milk over quite long distances.

Other changes during the depression are shown in Maps 4.38–40. The area of woodland (4.39) increased by some 27 per cent between 1873 and 1911, and the acreage of orchards (4.40) by about 21 per cent. The area of market gardens grew by an astonishing 145 per cent but even this is an underestimate of the extent to which farmers were cultivating vegetable field crops for human consumption. Fresh vegetables were not susceptible to foreign competition and although the distribution of market gardens is a result of many influences the dominant feature of Map 4.38 is proximity to large urban markets.

While grain acreages contracted from 1871 to 1911 yields per acre rose as poorer quality land was abandoned, whereas during the first half of the century yields increased more rapidly as acreages were expanding. Although sources are not as precise as we would wish it seems that average wheat yields rose from about 20 bushels per acre in England around 1800 to about 28 bushels in 1871 and 32 by 1911. Maps 4.41–48 provide a very rough estimate of the total output of wheat and barley, calculated by multiplying acreages by estimated yields. Output is expressed in bushels, a measure of volume, rather than by weight because of uncertainties in converting volumes to weight on a regional basis.

The final four maps (4.49–52) use an indicator of changes in the level of rents (Grigg, 1965). No data on actual rents exist in a form suitable for mapping but information is available on assessments for Schedule A of the income tax which were based on land rents. Levels of rent reflect many things: farmers' profit level, landlords' capital provision, types of farm enterprise, competition with other land uses, as well as the inherent quality of the land. It is difficult to generalize on a national scale; explanations for the changing fortunes of particular areas need to be couched in local terms. As an example the improvements of land in Norfolk and Suffolk are evident between 1806 and 1842 (4.49 and 50) while the depression after 1871 brought falling rents (4.51 and 52).

Maps make strict demands on data. Those presented here are as much a product of available sources as they are of an attempt to map the key processes of technological and institutional change in agriculture. The maps are not an end in themselves but a means to shed light on those processes and so deepen our understanding of agricultural change in the century and a half after 1750.

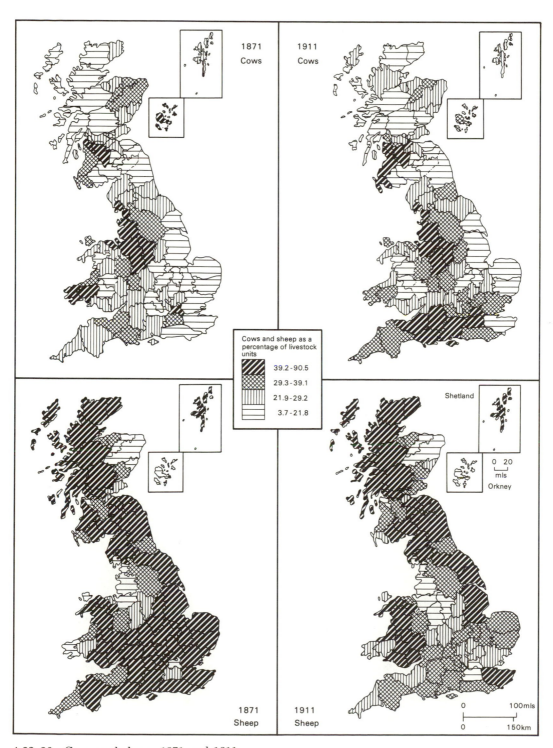

4.33–36   Cows and sheep, 1871 and 1911

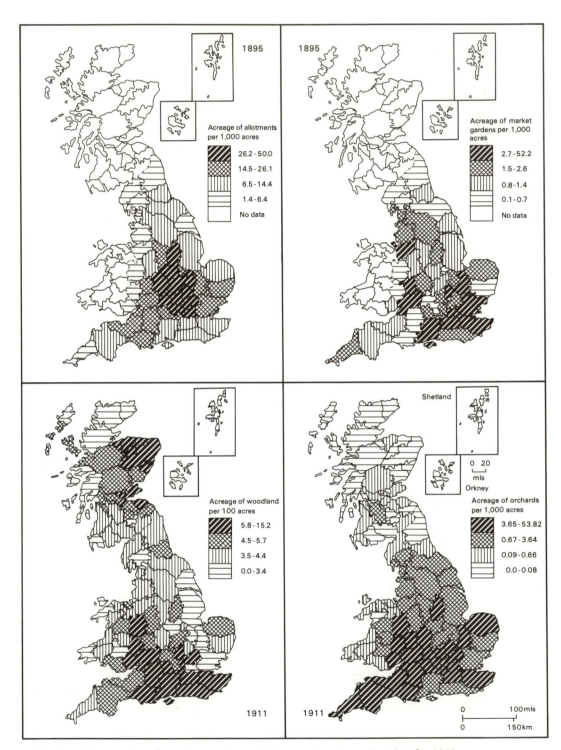

4.37–40 Allotments and market gardens, 1895; woodlands and orchards, 1911

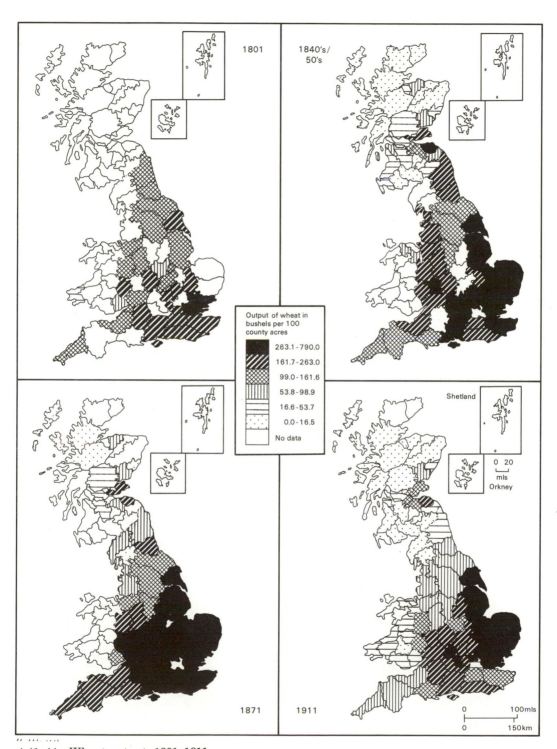

4.41–44 Wheat output, 1801–1911

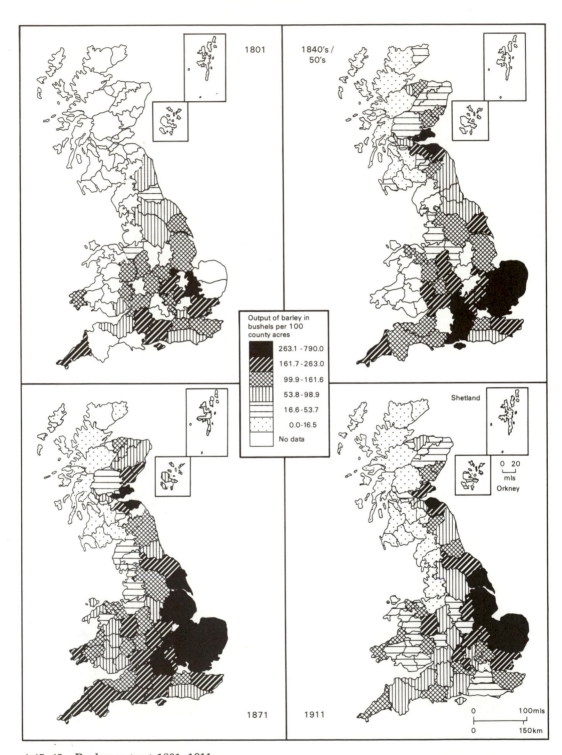

4.45–48 Barley output 1801–1911

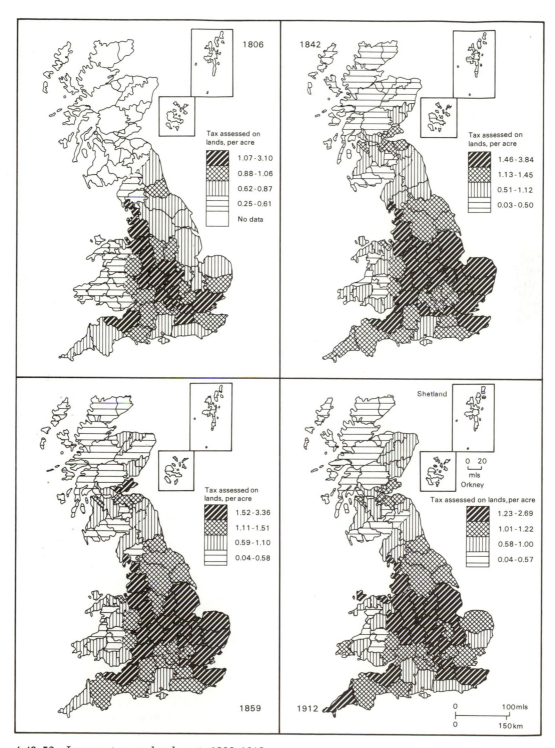

4.49–52 Income tax on land rent, 1806–1912

5 **Rural settlements**

Brian Roberts

The industrial revolution brought change to rural settlements, directly to those areas rich in resources, but indirectly to all regions as they were drawn more deeply into nationally organized systems of production. Map 5.1 conflates two imperfect sources: Thirsk's (1967) map of agricultural regions between 1500 and 1640 and Thorpe's (1964) map of rural settlement in 1962, for both reflect the deep-seated forces, physical, economic, social and historical, which created a multitude of local *pays*. As Thirsk (1967) has shown, these provided the varied contexts within which industrial genesis took place. All of the boxed details, irrespective of the processes of reproduction, are scale comparable. Tithe maps and Ordnance Survey maps (Harley, 1964; 1972) form a rich source for the study of nineteenth-century rural settlement, but while these have been used for numerous detailed studies it is regrettable that no attempt has been made to summarize the general picture. Work currently in progress may eventually rectify this omission.

Two general maps of regular village and composite plans suggest the presence of an underlying regional diversity, which ultimately has medieval roots (Roberts, 1979; 1982). These diverse regions, reflecting complex relationships between settlement types and environments, economies and societies, fall into eight categories for England and Wales as shown on the key to the central map 5.1. There was a clear though rough correlation between dispersed settlements in upland pastoral regions and nucleated village settlements in lowland open-field arable areas. Acklington and Grassington are representative nucleations, drawn from contrasting environments, while Armscote is characteristic of many haphazard agglomerations which may also have an origin in two or more small cells. In this latter case, the insertion of a sector showing open-field strips is a reminder that in the late eighteenth century many such villages experienced traumatic changes when large extents of townfield were enclosed and consolidated, and that farmsteads were often moving out to the new ring-fenced farms, bringing intercalated dispersion to landscapes once dominated by nucleation (Harris, 1961). A key problem in handling this material is whether to concentrate upon static description or to introduce the idea of change; both Acklington and Armscote contain a dynamic aspect, but the remaining illustrations are descriptive, showing contrasting types of settlement in wood pasture and open pasture regions (Thirsk, 1967). Although set within upland, Grassington is composite: the former cattle track, strongly marked by long, late, linear enclosures, once led to extensive common pasture still bearing the visible traces of prehistoric field systems. Narrow curved fields resulting from the piecemeal enclosure of strips, also seen in Wasdale Head, are typical of northern English uplands, as are the 'lathes' or scattered field barns. Llanwnda lies in Welsh lands, north of the Pembrokeshire 'landsker', where hill pastures intermingle with marginal stone-walled fields, while Bedevocle is a small, traditional loose textured Scottish 'fermtoun' (Dodgshon, 1980) which contrasts markedly with Wiston, a village in a part of Scotland more open to English influence since the middle ages (Parry and Slater, 1980). The plan of part of the Weald gives a glimpse of dispersion in wood pasture landscapes.

Changes to rural settlements during the nineteenth century included the restructuring of what had been inherited, intensification and colonization in zones of quickening economic activity, and depopulation in purely rural areas. In an important study Mills (1980), focusing upon socio-economic structure rather than morphology, has emphasized the presence of two types of village: those dominated by one landlord and one estate, and those in the hands of several or many proprietors, and while villages with divided or absentee landlords allow varied sub-types to be recognized, two basic lists of concomitant characteristics emerge. Some of the practical implications of these variations are illustrated by reference to Chambers's (1953) graph, based upon a sample of 119 villages in Nottinghamshire and Leicestershire which show some of the factors leading to settlement differentiation. The most powerful was the presence of industrial activity but amongst the purely rural villages the general rule is that the earlier the enclosure, the less marked the rise in population, and while no absolute correlation can be made with Mills's (1980) work, pre-parliamentary enclosure was inherently more probable in closed villages (Yelling, 1977). Social and economic differentiation are dominant themes in any view of changes brought by the industrial revolution.

Landownership forms one framework for change, and in Map 5.2 the English section is the most complete. Upon this background, derived from work by Clemenson (1982), Howell (1977), Timperley (1980) and ultimately Bateman (1876), has been superimposed the distribution of eighteenth- and nineteenth-century planned villages using studies by Darley (1978) and Lockhart (1980). These could either be remodellings, as at Acklington (5.1) and Revesby (mid-nineteenth century), or wholly new foundations. Tomintoul is one of a number of Scottish settlements planned between 1735 and 1850, while Easton Neston is representative of many villages planned in relationship to the great house. New accretive growth took place in old-established settlements, thus Grassington (5.1) lay in a lead-producing zone, but minerals also brought new villages, never truly rural, for they grew from the labour demands of industry. Throughout the coalfields and other industrial zones fields were turned over to crops of houses rather than grass (Jones, 1969), and the plans of East Howle in 1897 and 1970 are a reminder of the inherently ephemeral nature of these barracks for mineral extraction (Durham, 1951). The great slate quarries of Wales had a powerful impact upon the appearance of buildings the length and breadth of the land, but, as Barnes (1970) has shown, they and their like brought to inhospitable uplands the piecemeal intakes, quarrymen's smallholdings, and chapel clusters, as illustrated in the plan of a transect across a slate-producing area of Caernarvonshire (5.2, bottom left).

Changes in agriculture had repercussions on rural villages; new farm buildings were constructed to house new farming practices and new machinery, and cottages to house the labour (Loudon, 1839). In Ulgham for example (5.2) a remodelled late eighteenth-century farmstead has a circular structure attached, a horse gin, where two, three or more horses walked in an endless circle to work machinery in the barn. Only the main farm at the east end possessed that symbol of the new order, a steam engine.

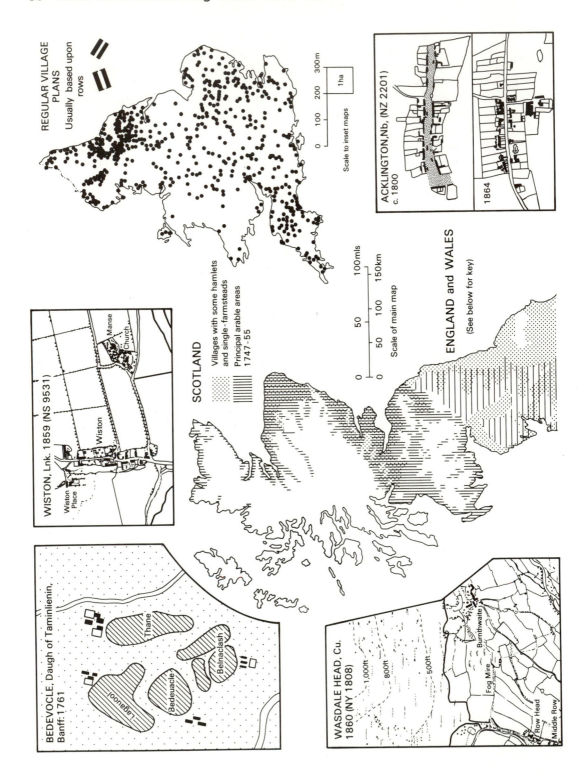

REGULAR VILLAGE PLANS
Usually based upon rows

ACKLINGTON,Nb, (NZ 2201)
c. 1800

1864

Scale to inset maps

0   100   200   300m

1ha

SCOTLAND

Villages with some hamlets
and single-farmsteads

Principal arable areas
1747-55

Scale of main map

0   50   100mls

0   50   100   150km

ENGLAND and WALES

(See below for key)

WISTON, Lnk. 1859 (NS 9531)

Manse
Church

Wiston

Wiston
Place

BEDEVOCLE, Daugh of Taminlienin,
Banff: 1761

Thane

Belnaclash

Bedeuacle

Lagaool

WASDALE HEAD, Cu.
1860 (NY 1808)

1,000ft

800ft

500ft

Burnthwaite

Fog Mire

Row Head

Middle Row

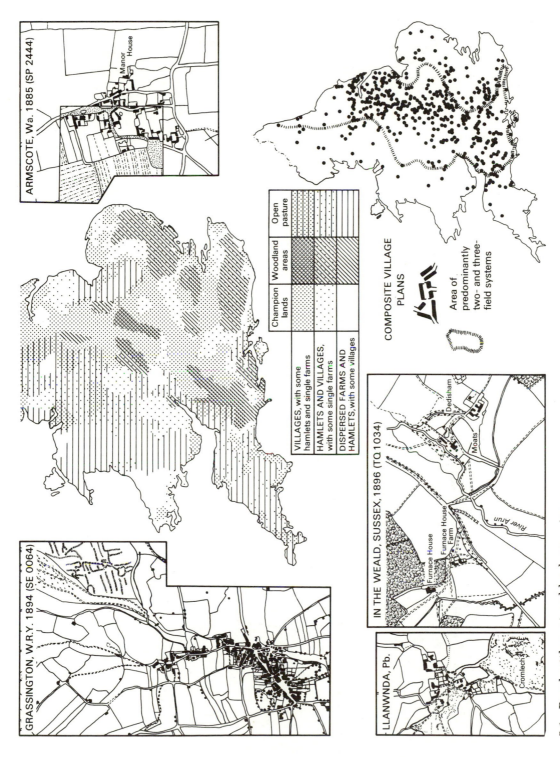

ARMSCOTE, Wa. 1885 (SP 2444)

Manor House

	Champion lands	Woodland areas	Open pasture
VILLAGES, with some hamlets and single farms			
HAMLETS AND VILLAGES, with some single farms			
DISPERSED FARMS AND HAMLETS, with some villages			

COMPOSITE VILLAGE PLANS

Area of predominantly two- and three-field systems

IN THE WEALD, SUSSEX, 1896 (TQ 1034)

Dedisham

Moats

Furnace House

Furnace House Farm

River Arun

GRASSINGTON, W.R.Y. 1894 (SE 0064)

LLANWNDA, Pb.

Cromlech?

5.1   Rural settlement and landscapes types

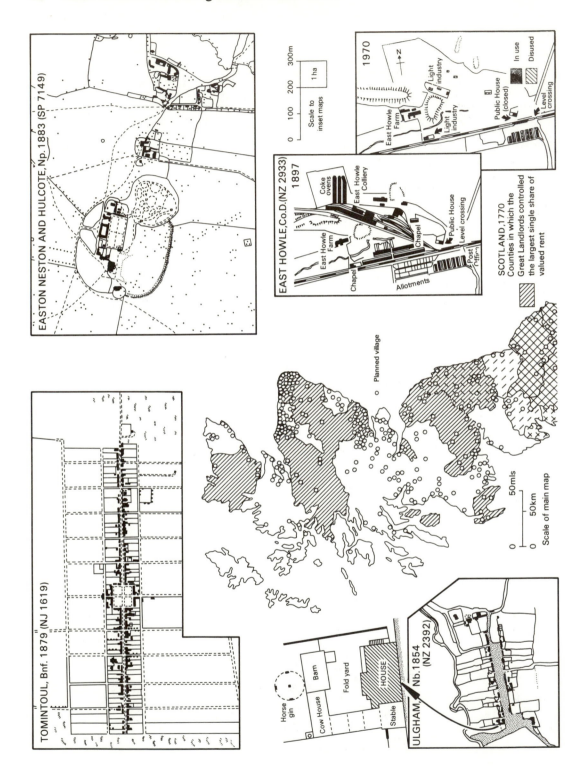

EASTON NESTON AND HULCOTE, Np. 1883 (SP 7149)

Scale to inset maps

0   100   200   300m

1 ha

1970

N

East Howle Farm

Light industry

Light industry

Public House (closed)

Level crossing

In use

Disused

EAST HOWLE, Co.D.(NZ 2933)   1897

Coke ovens

East Howle Colliery

East Howle Farm

Chapel

Chapel

Public House

Level crossing

Post Office

Allotments

SCOTLAND, 1770

Counties in which the Great Landlords controlled the largest single share of valued rent

Planned village

50mls

50km

0

Scale of main map

TOMINTOUL, Bnf. 1879 (NJ 1619)

Horse gin

Barn

Cow House

Fold yard

HOUSE

Stable

ULGHAM, Nb. 1854 (NZ 2392)

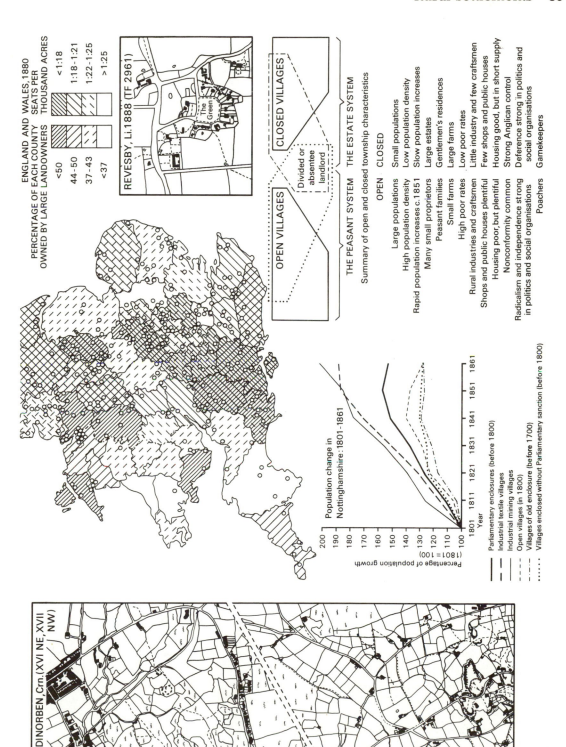

PERCENTAGE OF EACH COUNTY
OWNED BY LARGE LANDOWNERS

<50

44 - 50

37 - 43

<37

ENGLAND AND WALES, 1880
SEATS PER
THOUSAND ACRES

<1:18

1:18 - 1:21

1:22 - 1:25

>1:25

REVESBY, Li.1888 (TF 2961)

The Green

OPEN VILLAGES   CLOSED VILLAGES

Divided or
absentee
landlord

THE PEASANT SYSTEM   THE ESTATE SYSTEM

Summary of open and closed township characteristics

OPEN	CLOSED
Large populations	Small populations
High population density	Low population density
Rapid population increases c.1851	Slow population increases
Many small proprietors	Large estates
Peasant families	Gentlemen's residences
Small farms	Large farms
High poor rates	Low poor rates
Rural industries and craftsmen	Little industry and few craftsmen
Shops and public houses plentiful	Few shops and public houses
Housing poor, but plentiful	Housing good, but in short supply
Nonconformity common	Strong Anglican control
Radicalism and independence strong in politics and social organisations	Deference strong in politics and social organisations
Poachers	Gamekeepers

Population change in
Nottinghamshire: 1801-1861

Parliamentary enclosures (before 1800)
Industrial textile villages
Industrial mining villages
Open villages (in 1800)
Villages of old enclosure (before 1700)
Villages enclosed without Parliamentary sanction (before 1800)

200
190
180
170
160
150
140
130
120
110
100

Percentage of population growth
(1801 = 100)

1801 1811 1821 1831 1841 1851 1861
Year

DINORBEN, Crn.(XVI NE, XVII NW)

5.2 Planned villages and land ownership

6 Wives

E.H. Hunt

Wages were enormously influenced by industrialization. Longstanding differentials were swept aside and living standards changed in a variety of ways whose investigation forms the nucleus of that most contested of all economic history controversies: the debate on industrialization and working-class living standards. Most contributors to this debate have adopted a chronological perspective, assessing the influence of industrialization upon living standards mainly by measuring changes in real wages between the 1790s and the 1840s. Here the approach is spatial as well as chronological: it draws upon the techniques of both geographers and historians. And because industrialization started long before the 1790s, and was far from complete in 1850, the investigation begins in the 1760s and extends to the early years of the twentieth century. In fact, regional wage differentials can be described with more certainty than chronological changes in real wages in any particular locality: intensive industrialization occurred in relatively few places and its influence on the regional wage pattern was considerable. Moreover, when spatial analysis is combined with chronological analysis the effect of industrialization upon wages is more easily distinguished from the effects of demographic change and other influences. For these reasons a geographical approach to wages history can provide particularly valuable insights into the relationship between industrialization and living standards.

The starting point then is the wage pattern in the 1760s, when the founding members of Britain's factory proletariat began work at Richard Arkwright's Cromford Mill. Wages were then highest in the southeast of England, especially in and around London. As distance from the capital increased, especially to the north, wages fell. Map 6.1 shows carpenters' wages in c.1765, and at later dates, in six towns that have been selected to illustrate the changing pattern of wages over the industrial revolution period. Carpenters at Exeter received considerably less than London rates of pay but were marginally better-paid than carpenters in Manchester. North of the border wages appear to have been low by any standards: carpenters at Edinburgh earned about two-thirds, and those at Aberdeen about a half, of the Exeter rate. Map 6.2 supplements this evidence by showing in which English counties farm labourers were relatively well paid, or relatively badly paid, in 1767–70. The pattern is fairly clear: regional wage differentials were substantial and most of the highest-wage counties were in a broad belt in the southeast, extending from Hampshire to East Anglia. The greatest concentration of low-wage counties was in the north of England.

Unfortunately there are no surviving statistics of farm wages for Welsh and Scottish counties comparable to those shown for English counties on Map 6.2, although other evidence suggests that Welsh and Scottish farm wages in the 1760s were clearly below English rates. There are, not surprisingly, several other shortcomings in the evidence on regional wage differentials, not least the uncertain treatment of farm labourers' 'payments-in-kind' and the imperfect comparability of English and Scottish farm wages at those dates for which Scottish evidence exists. The evidence as a whole is discussed in detail elsewhere (Hunt, 1973; Hunt, 1986). Here it should be borne in mind that it would be unwise to place much reliance upon any individual wage reading shown on the map, and that differences in carpenters' real wages were a little less than the differences in money wages because rents tended to increase with town size. London rents were particularly high (22.21). Other spatial differences in the cost of living, however, were remarkably small and there is no evidence that regional variations in men's pay were

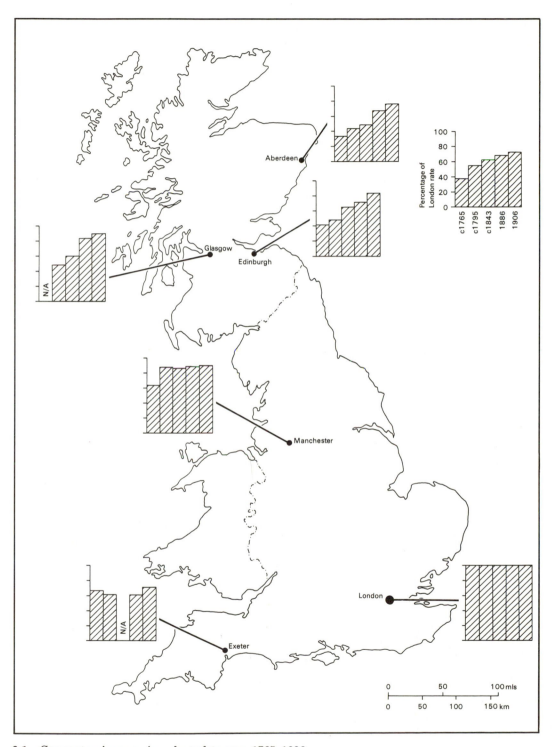

6.1   Carpenters' wages in selected towns, 1765–1906

much compensated by differences in the earnings of their wives and children. Fortunately, regional wage variations were so great that their overall pattern is beyond dispute despite the limitations of the wages evidence. The evidence of money wages is also corroborated by the migration of workers towards high-wage centres, by levels of county poor rates, and by consumption patterns that paralleled differences in money wages, in particular by the contrast between the wretched farm labourers in low-wage areas, who ate much bread and little meat, and their better-nourished counterparts elsewhere.

Between the 1760s and the 1790s the pattern of regional wage differentials was transformed. On Map 6.1 it is the improvement in Manchester wages that catches the eye. In three decades carpenters' wages there advanced from 64 per cent to 88 per cent of the London rate, and from a little below to well above the rates paid at Exeter. Carpenters' wages at Edinburgh and Aberdeen also moved closer to London rates, but their relative improvement was less than that at Manchester and left wages in the Scottish cities still well below those in Exeter. Lancashire is one of six counties selected to illustrate long-term changes in farm wages in different regions (6.3) and comparison with wages in Suffolk and Dorset shows that between the 1760s and 1790s Lancashire farm labourers enjoyed a relative improvement in wages similar to that of Manchester carpenters. Evidence for the two Scottish counties represented on Map 6.3 is available for the 1790s and suggests that farm labourers employed in the highland county of Ross and Cromarty must have lived badly indeed because their wages were only two-fifths of the rate in the highest-wage county and only two-thirds of the modest wages paid in lowland Dunbartonshire.

Map 6.4 includes Scottish as well as English counties and thus is not directly comparable with 6.2. It shows that Scottish farm wages in the 1790s were distinctly low when judged by English standards. There were a number of less badly paid counties in central and southern Scotland, but all of the twenty-one counties with farm wages of 6s. 6d. a week or less were north of the border. The most interesting feature on Map 6.4, however, is that most of the better-paid counties were now in the north of England, a marked contrast with the position in 1767–70. This change is most clearly evident when Maps 6.2 and 6.5 are considered together. At the earlier date none of the 11 highest-wage English counties were to the north of Nottinghamshire and most were in the southeast: by 1794–5, 6 out of 11 high-wage counties were north of Nottinghamshire, 2 others were in the Midlands, and only 3 were in the southeast. Five northern counties which appeared among the lowest-wage counties in 1767–70 (Northumberland, Westmorland, Lancashire, and the North and West Ridings of Yorkshire) were among the highest-wage counties by 1794–5. London remained a high-wage centre and wages were still high in some counties nearby. But southern counties more distant from London had suffered a substantial fall in relative wages.

In the nineteenth century the regional wage pattern changed considerably less than it had changed in the second half of the eighteenth century. The one notable change was the continued relative improvement in Scottish wages. This advance can be seen in Maps 6.1 and 6.3 and by comparing 6.4 and 6.6. By the 1840s, as compared with the 1790s, carpenters' pay in Glasgow, Edinburgh and Aberdeen had moved closer to rates paid in Manchester and London, although the gap was still considerable (6.1). By 1906, however, Glasgow carpenters' wages were the equivalent of carpenters' wages at Manchester and only 10 per cent short of the London rate. Aberdeen carpenters' pay had

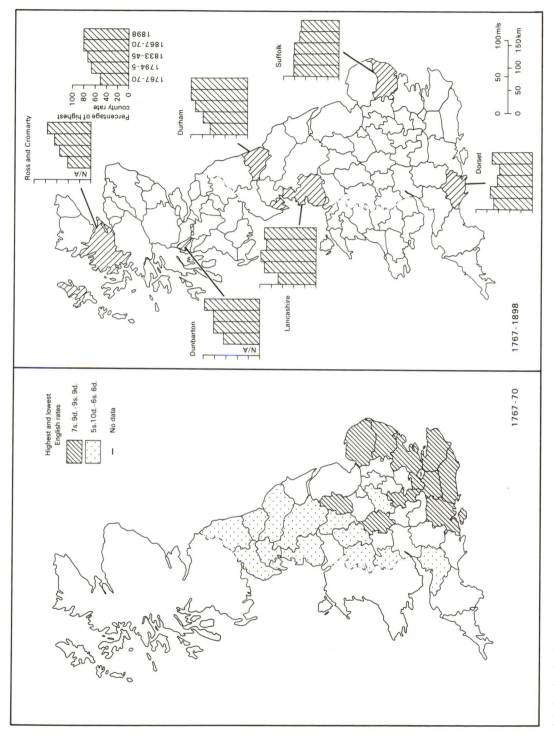

6.2–3   Agricultural labourers' wages, 1767–70 and 1767–1898

meanwhile caught up with, and passed, the rate paid at Exeter. Similarly with Scottish farm wages. In Ross and Cromarty in the 1790s, the 1860s, and 1898, wages were respectively two-fifths, two-thirds, and then three-quarters of the highest English county rates (6.3). By the last of these dates Ross and Cromarty labourers were better-paid than labourers in Dorset and Suffolk. Over the same period Dunbartonshire farm wages had improved from a level around two-thirds of the best English rates to among the highest in Britain. Central Scotland as a whole was one of Britain's high-wage areas by the end of the nineteenth century (6.6), and whereas in 1794–5 Scotland had accounted for all of Britain's lowest-wage counties, in 1898 17 of the 22 lowest-wage counties were in England or Wales (6.4, 6.6).

Changes in the spatial wages pattern also continued in some parts of England during the first half of the nineteenth century. This consolidation of the eighteenth-century transformation can be seen, for example, in the further advance of Durham wages between the 1790s and 1860s and in the relative fall in wages in Dorset and Suffolk (6.3). There was also, in the second half of the century, a trend in which the southeast of England (outside the London area) gradually displaced the southwest as the lowest-wage region (6.5, 6.7) and, by 1898, indications that south Wales was on the threshold of becoming a fourth high-wage region alongside the north of England, the London area and central Scotland (6.6). The outstanding feature of the wages geography of nineteenth-century England and Wales, however, is not these minor changes but the persistence throughout the century of the wage pattern already evident in 1800. Northern England was then relatively prosperous and the south (except around London) was relatively poor: this was still the position a century later. This persistence of the late eighteenth-century wage pattern and the contrast between the continuity of the nineteenth-century and the eighteenth-century transformation are both quite clear when Maps 6.2, 6.5 and 6.7 are examined together. The similarity between the wage patterns in 1794–5 and 1898 is as remarkable as the contrast between these patterns and that which existed in 1767–70.

The main features of the wages geography of Britain during the century and a half of industrialization are thus reasonably clear. Major regional differences in wages already existed at the beginning of the classic industrial revolution. Because industrial growth focused on parts of the north where wages had been low, industrialization initially eroded the pre-industrial pattern of wage differentials. Wage differentials then widened again towards the end of the eighteenth century, but in a quite different pattern from that which had existed in 1760. This new pattern was little changed throughout the nineteenth century, save for the continuing relative improvement in Scottish wages.

The pattern of changes in regional incomes likely to occur during industrialization has long been debated among development economists and economic geographers (Myrdal, 1957; Richardson, 1969). The traditional (neo-classical general equilibrium) interpretation emphasizes the probable erosive influence of labour migration and other market forces upon the initial income advantages of industrializing regions and predicts, therefore, that spatial income differentials created by industrialization are likely to be short-lived. A revisionist or 'growth pole' interpretation concedes some effect to these influences but denies that market forces are likely to erode regional income differentials rapidly. On the contrary, this school predicts that in several respects market forces may

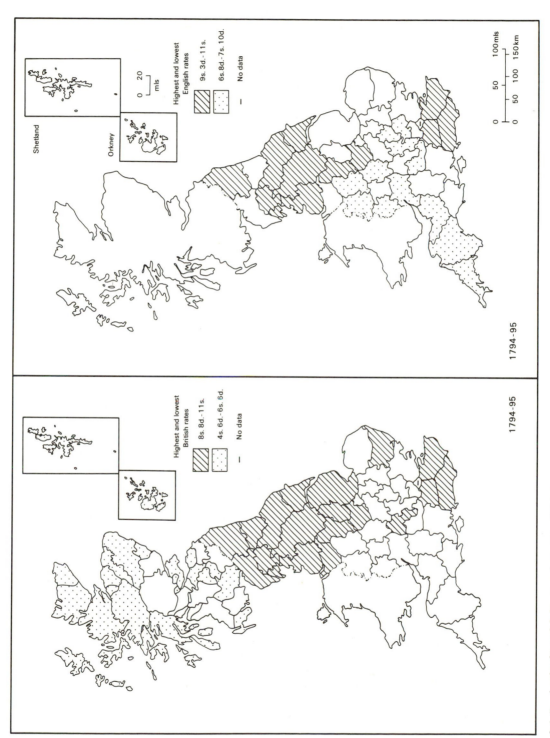

Shetland

Orkney

100 mls

50   100

50   100   150 km

0

0

1794 - 95

Highest and lowest
English rates

9s. 3d. - 11s.

6s. 8d. - 7s. 10d.

No data

0   20

mls

1794 - 95

Highest and lowest
British rates

8s. 8d. - 11s.

4s. 6d. - 6s. 6d.

No data

6.4–5 Agricultural labourers' wages, 1794–5

operate to reinforce the initial advantages of industrializing areas. The development of external economies and of economies of scale, it is argued, will encourage the clustering of new economic activities around the original industrial growth pole. In addition, rising incomes and rising productivity might stimulate local demand, and inmigration may contribute to this process by increasing local markets and by drawing disproportionately upon the more skilled and vigorous parts of surrounding populations. Moreover, the greater efficiency of up-to-date factories and workshops may well cause their products to outsell the more expensive articles produced by low-wage hand labour in peripheral regions. In short, the combination of 'clustering' effects at the growth pole and the unfavourable 'backwash' effects felt in peripheral regions may be sufficient to prolong and increase the initial advantages of industrialization (Richardson, 1969).

Which of these conflicting theories is more consistent with the changing pattern of British regional inequalities? Although industrialization at first reduced spatial wage differentials (which is certainly not consistent with growth pole theory), the British experience over the whole period 1760–1900 clearly affords far more support for the growth pole school than for the general equilibrium alternative. There was some erosion of regional wage variations towards the end of the nineteenth century (Hunt, 1973) and after 1914 inflation, followed by prolonged depression in the industrial districts, swept aside the spatial wage differences created by the industrial revolution and went some way towards restoring the pre-industrial pattern of regional prosperity. But erosion of wage differences had been hardly perceptible before the 1890s and the post-1914 transformation came far too late to diminish the broad consistency between the British experience and that predicted in growth pole theory.

What are the implications of these remarkable spatial differences in wages for the long-running debate on industrialization and working-class living standards (Lindert and Williamson, 1983)? There is obviously far more to 'living standards' than wages alone, but wages were their most important component and analysis of the effects of industrialization upon regional wage variations suggests that aspects of this relationship warrant more attention than they have received. One implication of what has been said is that enquiries into the course of real wages during industrialization are likely to produce very different answers according to which part of the country is investigated. Another is that anyone bold enough to generalize about real wage trends in Britain as a whole should take account of the effect upon the average of migration from lower- to higher-wage areas.

A possible reason why regional wage variations have been insufficiently considered in the great debate on living standards is perhaps evident in the earlier analysis of the rate of change in spatial differentials during industrialization. There was no dramatic change in the pattern of English differentials between the 1790s and 1840s, the years the debate has focused on: the main changes had occurred earlier. But this observation suggests another reason for dissatisfaction with the way the debate has proceeded; the effect of industrialization upon workers' living standards is the central issue but (for reasons not entirely clear) most commentators have regarded the initial decades of industrialization, when its effects upon the regional wage pattern were greatest, as chronologically peripheral. And if attention to previously neglected decades seems likely to reduce the general inconclusiveness of the controversy, similar beneficial

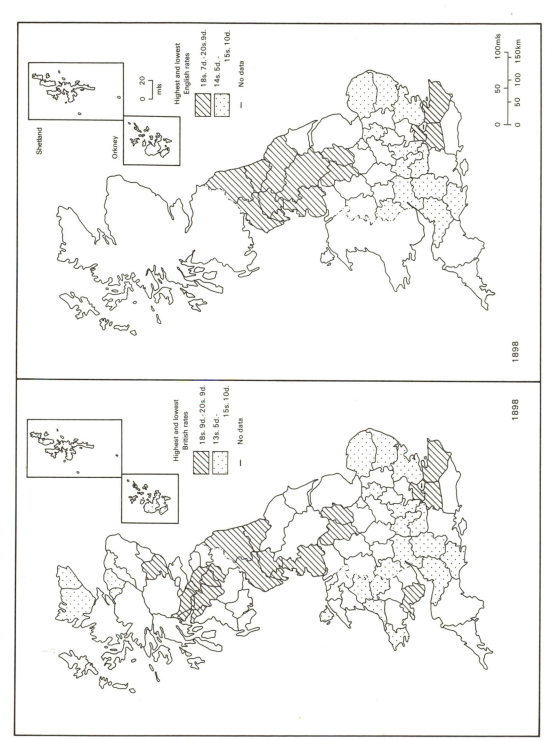

Shetland

Orkney

Highest and lowest
English rates

18s. 7d.–20s. 9d.

14s. 5d.–

15s. 10d.

— No data

0 20
mls

100mls
150km

0 50
0 50 100
0 50 100 150km

1898

Highest and lowest
British rates

18s. 9d.–20s. 9d.

13s. 5d.–

15s. 10d.

— No data

1898

6.6–7 Agricultural labourers' wages, 1898

results might well repay attention to traditionally neglected areas of Britain. Real wages in Scotland call for particular attention, as, on the evidence examined here, they appear to have been characterized by long-term improvement.

The effect upon average wages of migration towards high-wage areas, the extent and direction of changes in regional wage differentials in the second half of the eighteenth century, the evidence of Scottish wage trends, and the general association between industrialization and high wages that is immediately evident when chronological wage-analysis is supplemented by spatial analysis, each add weight to the more optimistic interpretations of the effect of industrialization upon workers' living standards. As emphasized above, none of the surviving wage statistics will bear much weight of analysis, but the long-term trends of wages in Lancashire and in a southern county such as Buckinghamshire serve to summarize the most obvious implication of regional wage differences (Table 6.1).

Table 6.1 Farm labourers' wages, 1767–1898: Lancashire and Buckinghamshire

	1767–70	1794–5	1833–45	1867–70	1898
Lancashire	6s. 6d.	10s. 1d.	12s. 5d.	17s. 9d.	19s. 4d.
Buckinghamshire	8s. 0d.	7s. 4d.	9s. 10d.	14s. 3d.	15s. 2d.

Sources: See p. 229.

In 1767–70, when industrialization was beginning, Buckinghamshire farm wages were over 20 per cent above farm wages in Lancashire. As Lancashire industrialized, wages there rose rapidly until, by 1794–5, they were more than a third above the level in Buckinghamshire. This proportionate difference was not subsequently increased but Lancashire retained an advantage of between a quarter and a third over Buckingham-shire wages to the end of the nineteenth century. Higher wages were reflected in better dress and diet, in lower levels of pauperism, and in other aspects of working-class well-being. None would deny that industrialization was the main cause of Lancashire's advantages. And most probably would agree that Buckinghamshire provides a good example of the plight of peripheral regions that experienced population increase unaccompanied by industrialization. From this perspective the favourable effects of industrialization upon living standards are indisputable.

7 Wind and water power

Paul Laxton

Before the successful application of steam power to rotary motion in the late 1780s, most industrial processes either relied upon intensive human labour or were carried out with the aid of animals, wind or water. The horse gin was applied to textile production, for winding purposes in mines, and in a wide range of crushing processes, especially in agriculture. Windmills, although chiefly used for flour milling and water pumping, were widely applied to industry both in low-lying areas with insufficient water power potential and elsewhere where demand dictated. Waterwheels, on the other hand, were widely applied to industrial tasks and were crucial to the early stages of industrialization. They drove yarn-spinning mills, fulling stocks, tilt hammers, furnace bellows, ore crushers, grinding wheels and machines for the manufacture of dyes, needles, gunpowder, paper and leather. Their oldest application was in processing farm products including flour milling. Eighteenth-century experiments greatly increased the efficiency of such applications and widened the scope for both windmills and watermills. In some industries it was several decades before steam was regarded as superior to these earlier sources of power.

It is easier to appreciate the ubiquitous application of wind and water power than to chart its geography. From the 1830s, the Factory Inspectors published returns of the number and power of waterwheels in textile factories. In 1838, waterwheels provided 21 per cent of the power in British cotton mills and 43 per cent for woollen mills. By that time water power was significant chiefly in the peripheral textile areas; for the earlier years when it was crucial to industrial development we have no such data. Maps 7.1–4 have therefore been based upon large-scale county maps whose makers generally plotted the sites of windmills and watermills. They give a good general impression of the distribution of such sites but have significant limitations. They are undoubtedly incomplete. Some 5 per cent of the sites might be missing in the cases shown here. Those maps involving more than one county are based upon surveys conducted at different dates by different surveyors. Not all the mills shown in East Anglia were necessarily operating simultaneously. General practice was to show a set of wheels with one symbol, so that the cartographer's symbol may indicate one, two or more wheels. The mapmakers gave no systematic indication of the use to which the power was put and we have no means of estimating the horsepower at each site. Local studies have tended merely to describe surviving sites and have made little contribution to the systematic evaluation of these data.

The most important underlying influence on the regional concentrations of wind and water power is not natural resources or environmental conditions but the critical nature of local demand. True, some parts of the country were too flat to provide adequate fall for watermills (7.2), but for every stream upon which for practical purposes there was no scope for further waterwheels there was another, often close by, whose power potential was unrealized (7.1, 3 and 4). We should ask therefore not why industry in the late eighteenth century located in areas of abundant water power, but why it failed to locate in areas of equal potential as wind and water power extended well into the steam age. The greatest concentrations of waterwheels, such as those around Stroud (7.1) or Sheffield (7.3, inset) can no more be explained in terms of physical geography than the concentrations of windmills that had clustered around medieval towns and continued to characterize many eighteenth-century towns. Of the 109 windmills in Lancashire in the late 1780s, twenty-four were within a mile of Liverpool Town Hall.

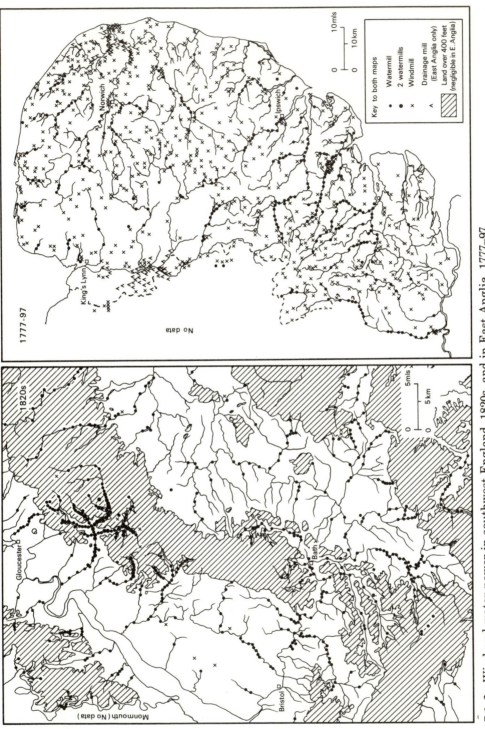

7.1–2   Wind and water power in southwest England, 1820s, and in East Anglia, 1777–97

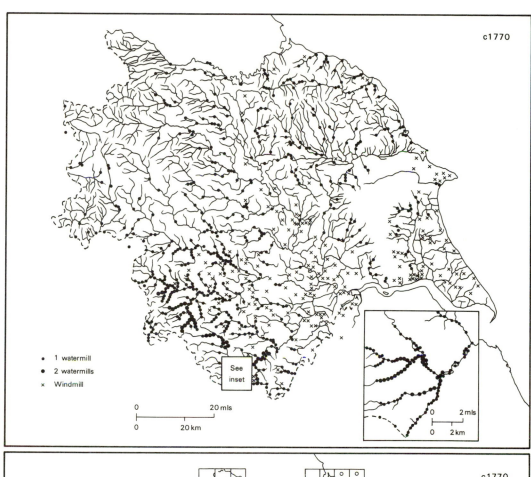

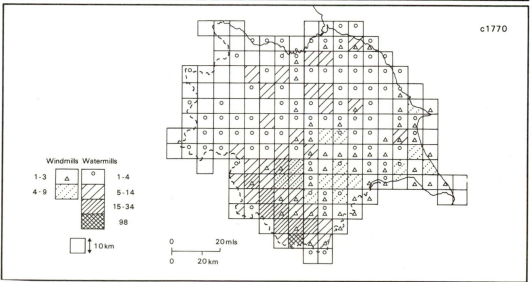

7.3–4 Wind and water power in Yorkshire, *c.* 1770

8 Coal and steam power

Nick von Tunzelmann

Maps 8.1–4 show the output of coal by region. The detailed geography of the coalfields is mapped on 1.3. Only the substantial fields would have had a detectable influence on prices except at the local level. The precise distribution of pits within coalfields at this time is rarely known with much confidence. For instance, the proportion of the South Wales coalfield worked intensively was quite small. The bibliography of M.W. Flinn's recent authoritative volume of the National Coal Board's History Project includes many studies of individual fields (Flinn, 1984). The same source has revised the previously accepted estimates of output by coalfield over the years 1700–1830, and the revisions are included in the maps.

The reliability of the earlier output figures is obviously low, though the NCB Project has done much to put all coal output data on a consistent footing. Flinn's figures are about 30 per cent higher in total than Pollard's recent estimate (Pollard, 1980). Annual figures are obtainable from official sources from 1854 onwards. Robert Hunt, the first Keeper of Mining Records, noted his procedure for compilation in a letter written in 1854:

> The statistics of coal and iron have been obtained by visiting the coal and iron producing districts, and by issuing circulars of inquiry. The replies given to these (non-compulsory) applications, whether personal or otherwise, have been full and satisfactory. The information thus obtained has been corrected by ascertaining the quantities exported and carried by railway and canal from each district; and by computing, after careful inquiries made on the spot, the local consumption. In all cases the railway companies have given to this office exact returns of their mineral traffic. (*BPP* 1856: LV, p.469)

Mitchell and Deane (1962) publish an annual series (with a slightly different regional breakdown, and including Ireland), and are thus led to doubt the quality of the series before 1873; but by the standards of the period it is probably fairly accurate.

Coal is such a bulky commodity that transportation costs rather than extraction costs tended to dictate the economy of using coal in the industrial revolution period. Coastal shipping of coal was well-established by the eighteenth century, and prices of sea-coal did not differ so greatly among the ports. In the early 1840s, most seaports not especially favoured by proximity to coalfields would have been able to market the coal at between 16 and 22 shillings a ton. Thus much of the coast even of the Scottish Highlands and Islands could get coal for under £1 a ton, while the ports of southern and eastern England clustered around this figure. Overland transportation was a very different matter; haulage by waggon in the days preceding the railway era was very expensive. A common assertion (Flinn, 1984) ran that the price of coal would be doubled by carriage of just 10 miles from the coal-pit. Assuming a typical pit-head price of 5 shillings a ton, this would imply a charge of 6*d.* (2.5p) per ton-mile – a figure supported by other evidence (Jackman, 1916). The contours for prices of 10 and 20 shillings per ton in Map 8.5 reflect these factors, of low per-mile cost of coastal shipping and high per-mile cost of waggon traffic. The data for England and Wales are returns for 480 Poor-Law Unions (averaging 1842 and 1843) (von Tunzelmann, 1978) (*BPP* 1843: XLV). The data for Scotland are even more profuse, being returns from 650 localities, mainly relating to 1842 (*BPP* 1844: XXIII); the returns have been used rather differently by Levitt and Smout (1979). The potentialities of such sources for analysing regional price differences, not only in regard to coal but for almost all items of staple consumption, have scarcely been touched. The

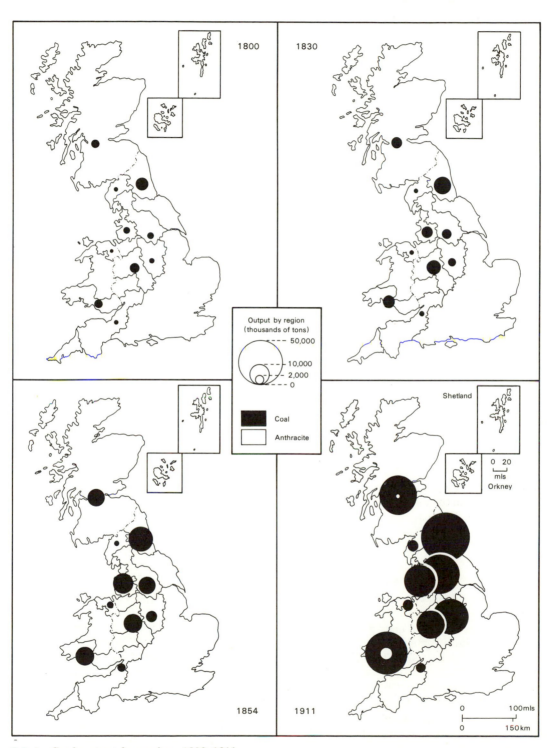

8.1–4   Coal output by region, 1800–1911

Journal of the Statistical Society, for example, regularly published regional price data through the 1840s.

Now the prices so quoted are for household coal, whereas that utilized by industry may have differed in quality and price. Steam engines at collieries, for instance, generally burned otherwise unsaleable 'slack', and it was primarily for this reason that many persisted with older 'atmospheric' engines of the Newcomen type long after the advent of the Watt engine (von Tunzelmann, 1978). By itself, this widened the regional disparities. On the other hand, in areas where coal was extremely expensive there may have been the alternative of other fuels. In the more inland parts of the Scottish Highlands where coal could rise to £1.50 or £2 a ton it was sometimes stated that only the rich purchased coal at such prices. Common substitutes both there and in the Scottish Lowlands were wood, moss and peat – the return for Fodderty in Ross and Cromarty for example speaking of 'a stock of peats, 6 foot square by 7 foot high, 20 shillings'. Peat was sometimes burned in steam engines in Cornwall and Devon.

Regional price differentials in 1912, based on a much smaller sample, had greatly narrowed (8.6), demonstrating of course the impact of the railway. The contrast between internal and seaborne transportation before the railway age had, if anything, been reversed. South-coast ports like Southampton and Dover emerge as the most expensive. Obviously general price movements restrict comparisons between 1842/3 and 1912. The average pit-head price of coal in England in 1912 was 8s. 9d., in Wales 11s. 5d. (including anthracite coal), and in Scotland 6s. $9\frac{1}{2}d$., according to the relevant Mineral Statistics. The year 1912 was quite an expensive one for its time, partly as the result of major strikes. These prices compare with the customary 4s. to 7s. in 1842. The data come from the Board of Trade Enquiry into the regional pattern of working-class cost of living (Cd 6955). A similar enquiry for 1905 summarized in the same document yielded a very similar regional pattern, at prices about one-fifth lower.

The most thorough attempts to trace the diffusion of the steam engine in the eighteenth century have traced 2191 engines (8.7). The casual evidence of contemporaries on the number of engines in specified areas suggests a possible total of 2400 to 2500 engines built in the eighteenth century (Kanefsky and Robey, 1980). This upward revision of previous counts follows a long tradition, but it should be noted that there are probably some upward as well as downward biases in these data. Twenty-seven of the engines were 'experimental', i.e. not applied to industrial purposes. The figures relate to the number actually standing in 1800, but it is difficult to eliminate the double counting of engines moved from one site to another. Kanefsky and Robey counted 478 Boulton & Watt engines erected in Great Britain, which Jennifer Tann's recent and definitive count reduced to 451 (1981). Most of the engines for which Kanefsky and Robey have not been able to prepare computer cards were probably small, e.g. winding engines on the Staffordshire coalfield.

While most effort has gone into counting the number of engines, the really crucial factor for gauging their impact was their power. Kanefsky and Robey unearthed horsepower or cylinder-size data for 607 of their 2191 engines, but these are dominated by Watt engines (434 engines at an average horsepower of 26.6; the average for the 173 non-Watt engines was 24 horsepower). Extrapolating these averages would give total horsepower built by 1800 of over 50,000, but undoubtedly the 'averages' are heavily biased towards larger engines – Kanefsky elsewhere (1979) estimates total steam horsepower of about 35,000 in 1800. The county-level horsepower figures in Map 8.8 are

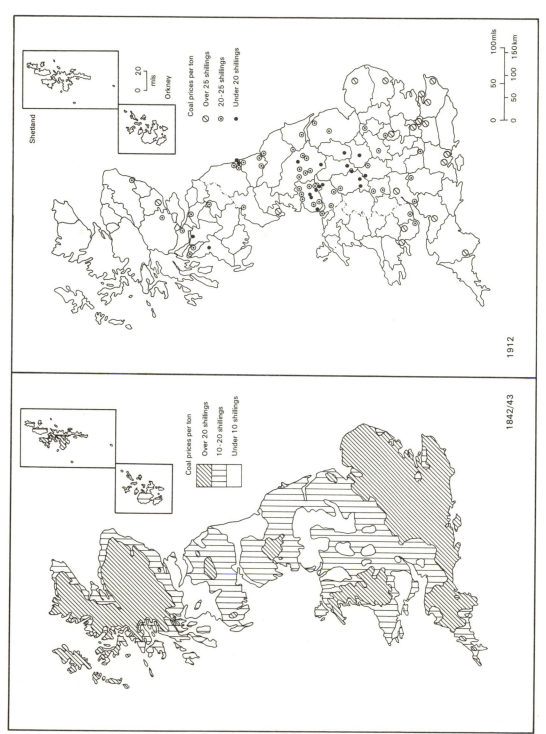

Coal prices per ton
⊘ Over 25 shillings
⊙ 20–25 shillings
● Under 20 shillings

Shetland

Orkney

0   20
mls

0   50   100mls
0   50   100   150km

1912

Coal prices per ton
▨ Over 20 shillings
▤ 10–20 shillings
☐ Under 10 shillings

1842/43

8.5–6 Coal prices, 1842/3 and 1912

for Boulton & Watt engines only, taken from von Tunzelmann (1978, p. 148). They make a small allowance for double-acting engines by simply doubling the nominal horsepower from single-acting engines. Evidence subsequently encountered suggests this to be a slight overstatement. Some county-level differences in the power rather than the number of engines are apparent from the map, even allowing for the limitation to Watt engines. Cornwall has easily the largest aggregate horsepower in Watt engines (and probably in all engines), whereas it comes only fourth in Kanefsky and Robey's count of number of engines.

Data on the number of engines and their horsepower in the nineteenth century are much weaker still than those for the preceding century, despite some reasonably well-founded data for some particular industries – estimates such as those of Mulhall (1903) that one sometimes encounters must be treated as sheer guesswork. An especially serious problem over the years between about 1850 and 1880 is correctly assessing power output. Until the middle of the century the conventional procedure for estimating the horsepower of an engine was, following James Watt, relating it to the dimensions of the engine's cylinder. With the spread of the practices of compounding cylinders and expanding the steam in the cylinder, this 'nominal' horsepower became an increasingly serious underestimate of power output – the more valid measure of 'indicated' horsepower later took over, but alas more slowly than the spread of the new practices themselves. This deficiency bedevils some recent assessments of nineteenth-century aggregate horsepower, especially that by Musson (1976), with its substantial reliance on the 1870 Factory Returns. Kanefsky (1979), in a generally convincing criticism of Musson's survey, points to: (i) this confusion between nominal and indicated horsepower; (ii) incomplete returns from local authorities to the circulars of inquiry (only eight English counties sent full returns of workshops); (iii) incomplete coverage of the Workshops Act of 1867, which was limited to workshops employing female or child labour; (iv) inexperience of the factory inspectors in dealing with the whole range of manufacturing industries outside textiles which the Extension Acts of 1864 and 1867 opened up to them. Casual comparisons with the 1871 Report on Coal Supply and the 1871 Population Census suggest to Kanefsky that the under-reporting of steam power in manufacturing may run to some 15 per cent, apart from the complete exclusion of some activities nowadays classified as manufacturing, of which the most important was grain milling. These limitations have to be stressed in considering Map 8.9 showing steam power in manufacturing by county. Non-manufacturing usage of steam power has been excluded.

The first Census of Production in 1907 had no regional breakdown, though for a majority of industries steam horsepower (distinguishing steam engines from steam turbines), as well as other primary power sources, is reported separately for Scotland, Ireland, and England and Wales. In Table 8.1 I have attempted to complete the division between Scotland and England and Wales, utilizing the 1911 Occupational Census and the 1924 Census of Production.

There is an important difference between the 1870 estimates on the one hand and those for 1907 and 1800 (subject to previously mentioned qualifications) on the other, in that the latter refer to total installed capacity, whereas the 1870 Returns attempt to gauge power actually in use. Not till the 1924 Census do we have reasonable reliable estimates of both power-in-use and power capacity at the same time. Kanefsky's (1979) guesstimates of power capacity in 1870 (with his reckonings of power-in-use shown in

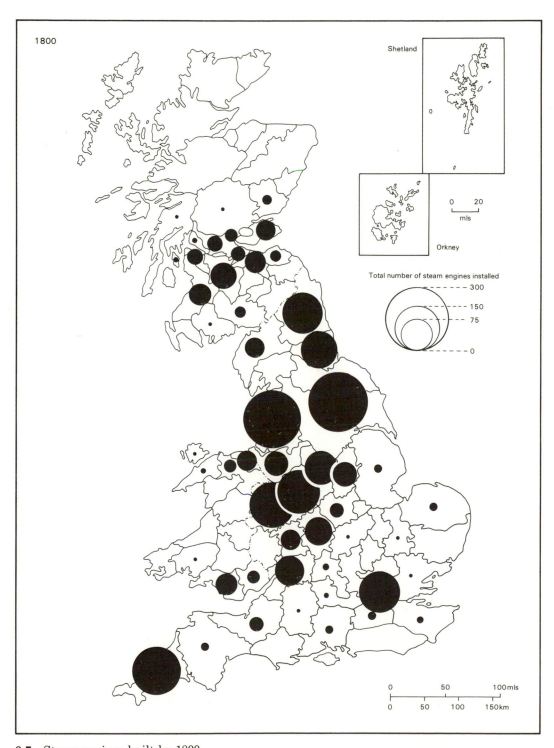

1800

Shetland

Orkney

0   20
mls

Total number of steam engines installed

300
150
75
0

0   50   100 mls
0   50   100   150 km

8.7 Steam engines built by 1800

parentheses) are as follows: Factory Return trades 1,500,000 (1,200,000), coal mines 500,000 (300,000), other mines 100,000 (60,000), waterworks and gasworks 60,000 (36,000), grain mills 90,000 (54,000), other 30,000 (18,000); total 2,280,000 (1,668,000). These figures are said to be for Britain but are probably for the UK (i.e. including the whole of Ireland).

In Table 8.1 steam power has been classified according to industry of use, with very rough approximations to present-day schemes of industrial classification (Industrial Orders). No adjustment has been made for under-reporting. No figures were given for agriculture but a Board of Agriculture survey estimated 106,000 hp on British farms in 1908. To repeat: the 1800 figures are of numbers of steam engines, those for 1870 are of horsepower in use, and those for 1907 are of horsepower capacity.

Table 8.1 Steam power in Britain in industry

Industry	1800		1870 ('000 hp)		Total %	1907 ('000 hp)				Total %
	No.	%	England Scotland & Wales			England & Wales (a) (b)		Scotland (a) (b)		
Mines & quarries	1064	48.6	360*	—	26.4	1924	3	453	0.5	25.0
Food, drink, tob.	112	5.1	66*	—	4.8	206	0.6	34	0	2.5
Coal & petr. prods	18	0.8	12	1	1.0	34	0	3	0	0.4
Chemicals	18	0.8				162	0.7	15	0	1.9
Metal manufacture			224	22	18.0	1179	28	240	6	15.3
Mechanical eng.	1263	12.0	40	5	3.3	227	6	37	0.2	2.9
Instrument eng.			0.1	0	0	1	0	0.1	0	0
Shipbuilding	0	0	5	5	0.7	104	0.2	34	0.5	1.5
Vehicles	0	0	1	0	0.1	31	0.7	3	0	0.4
Metal goods n.e.s.	**	**	24	1	1.8	77	0	5	0	0.9
Textiles	469	21.4	417	69	35.6	1639	13	165	0.6	19.1
Leather, fur	0	0	1	0.1	0.1	17	0	2	0	0.2
Clothing, footwear	0	0	1	0.1	0.1	15	0.7	2	0	0.2
Building materials	12	0.5	14	0.8	1.1	192	6	23	0	2.3
Timber, furn.	0	0	0.8	0.3	0.1	82	0.5	21	0.3	1.1
Paper, printing	13	0.6	26	7	2.4	129	0.2	44	3	1.9
Other manufactures	0	0	5	1	0.5	28	0	3	0	0.3
Gas, elect., water	36	1.6	36*	—	2.6	1167	382	117	38	17.9
Construction	0	0	1***	0.3***	0.1	317	29	25	0	3.9
Transport, etc.	52	2.4	118*	—	1.3	181	0.8	16	0	2.1
Other	43	2.0				21	0.1	5	0.9	0.3
Not known	89	4.1	—	—	—	—	—	—	—	—
Total										
Manufacturing	906	41.4	836*	115	69.6	4124	57	631	11	50.7
	2191	100	1366*	—	100	7734	472	1248	51	100

Sources: *1800:* Kanefsky and Robey (1980); *1870:* Kanefsky (1979) and *BPP* 1871: LXII; *1907:* Author's calculations from 1907 Census of Production
Notes: (a) steam engines; (b) steam turbines
    *   includes some or all Scotland
    **   included in metal manufacture, etc.
    *** very incomplete (remainder in *Other* category)

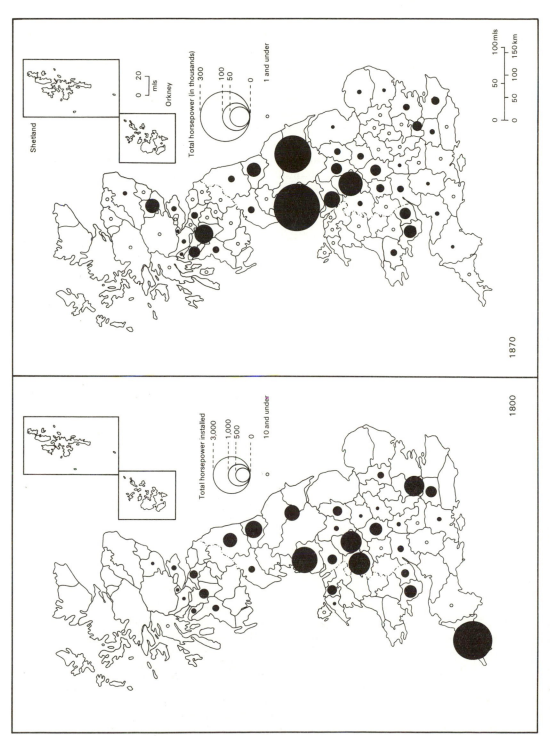

8.8–9 Boulton & Watt engines, 1800, and steam power in manufacturing, 1870

9 **Transport**
Michael Freeman

Road transport never symbolized the industrial transformation as canals and railways did. But for facets of economic activity where production was areal, as in agriculture, or where enterprise was widely diffused and small in scale, as in much early mining and manufacture, road transport enjoyed an importance of its own (Aldcroft and Freeman, 1983). In passenger travel, its position was not far short of monopoly. The principal agency of road improvement was the turnpike trust whereby Parliament granted temporary powers for groups of local people (trustees) to charge tolls on defined stretches of road and spend the income upon maintaining and upgrading them. By the 1830s some 22,000 miles, roughly one-sixth of the country's roads, were under trust administration (Albert, 1972). Map 9.1 shows the turnpike system of England and Wales by 1770, covering some 15,000 miles (based on Pawson, 1977). Most turnpikes were established in response to rising traffic levels so that the density pattern is reflective of the geography of economic growth up to 1770. Thus areas such as the West Midlands, Forest of Dean, the West Country woollen districts and the Weald stand out, whereas Lancashire and Yorkshire do not. London is revealed as a pivot of the system, a place held since at least 1750 and indicative of the capital's penetration into national economic life.

The condition of turnpikes varied greatly. In the nineteenth century, under Telford, Metcalfe and the McAdams, some uniformity of maintenance prevailed. Otherwise, standards varied as much as underlying soil and geology. The mere existence of a turnpike trust was of limited import. Turnpikes were not indivisible entities like inland navigations and railways. Faced with an impassable section the road traveller simply made a detour. And again unlike railway and navigation proprietors, trustees had no direct concern in the provision of transport capacity. Hence pressures for efficiency were largely self-imposed. The trusts were an administrative not a technical innovation. They must be assessed as such, with close reference to the local machinery of their operation.

That turnpike trusts achieved an aggregate improvement in roads is unquestioned. It was manifest in the scale and efficiency of public transport operations by the 1820s. Map 9.2 shows London's stage-coach connections by 1821. There were some 4500 separate departures each week giving a potential capacity of roughly 58,500 passengers. From 1750 onwards all coach schedules had been radically improved (9.3). The institution of regular horse changes *en route* and better vehicle design helped to reduce travel time; but so did better road surfaces. The degree to which services directly penetrated the provinces is an obvious measure of national integration. The strength of the Irish connections, via Milford, Holyhead and Liverpool, is especially revealing. But there is a very evident under-representation for northern industrial areas. This may have been a result of Birmingham operating as an interchange and break-of-bulk point for the north. It is as likely to have reflected the increased control those areas were registering over their own destinies – against, for example, the former dominance of London-based broking and merchanting.

Public carriage of goods by road had a longer pedigree than that of people. London possessed a national array of service links by at least 1637. Provincial-based operations date from at least the eighteenth century. Map 9.4 depicts the carriers operating from Leeds in *c.* 1827. There were few eastward links because effective capacity had long been provided over the Aire and Calder navigations. Services were concentrated in an arc

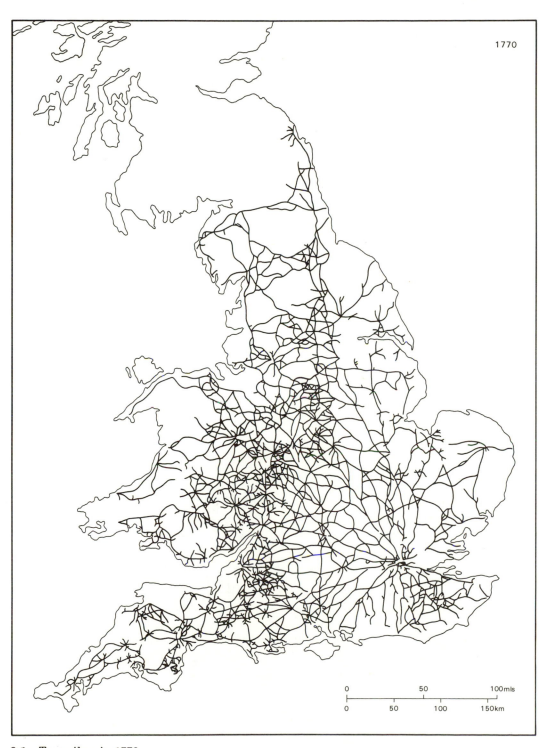

1770

9.1   Turnpikes in 1770

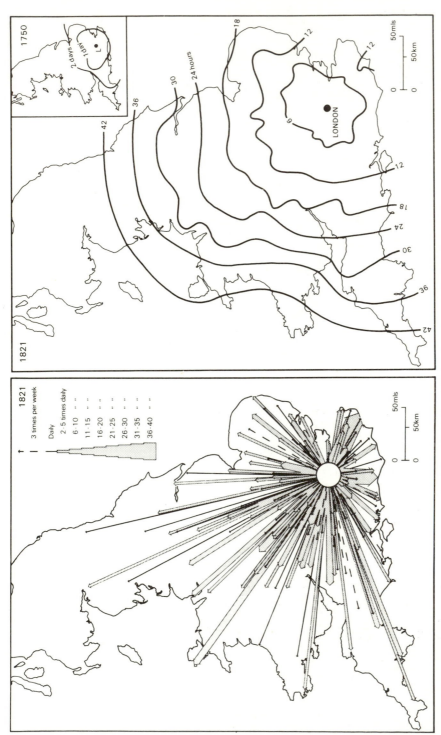

9.2–3   Stage-coach services from London in 1821

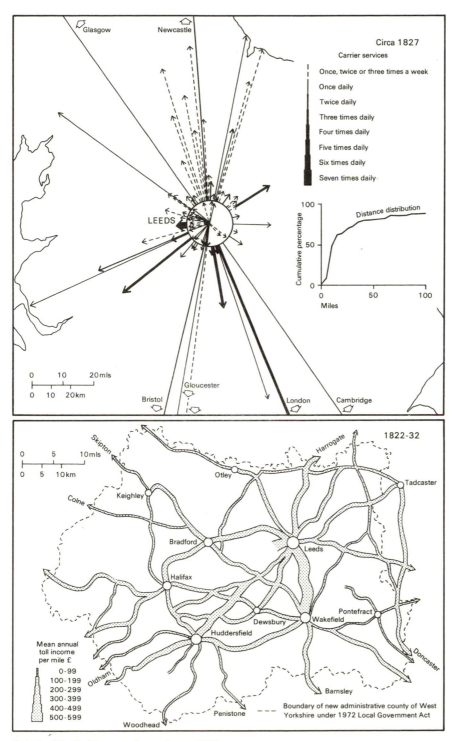

9.4–5 Yorkshire road carriage, 1820s

extending from Huddersfield in the southeast to Skipton in the northeast and thence out into the Vale of York. In the one case they serviced wool textile manufacture in all the complex spatial structure of its vertical integration; in the other the manufacturing population's daily requirements of sustenance. The distance distribution of carrier connections shows a clear bias towards short-haul movement: roughly 60 per cent fell within 10 miles. It should not be assumed, however, that this represented the distance distribution of commodity flow, for the onward transmission of goods by separate carriers was a regular practice. What is important is that Leeds was a ready point of access to a system of goods transport which penetrated country-wide.

The great lacuna in our knowledge of road transport lies in what can be established of traffic in general as distinct from scheduled operations. Map 9.5 depicts mean annual toll income per mile for turnpike trusts within the broad textile-working area of the West Riding of Yorkshire over the period 1822–32. Comprehensive income figures are available only after 1822 when trusts were first required to submit annual accounts to the local Clerk of the Peace. Interpretation of such a map requires utmost caution. Tempting as it may be, trusts' toll income is not a straightfoward surrogate for turnpike traffic. Variation in the number of a trust's tollgates can be fairly reliably offset by expressing income on a per mile basis, as already done here. But it cannot be assumed that levels of charged tolls were uniform between trusts. There is also a serious disaggregation problem in using data on trusts' income rather than income figures for separate gates. However, none of these interpretative difficulties can explain away the evident cohesion of the manufacturing area bounded by the five principal textile towns. Alongside such important thoroughfares as the Great North Road, it stands as a beacon of activity, perhaps testament to the rising fortunes of wool textile manufacture after the mid-1820s.

Most modern observers would agree with John Aikin's comment in 1795:

> The prodigious extensions made within a few years to the system of inland navigation . . . cannot but impress the mind with magnificent ideas of the opulence, the spirit and the enlarged views which characterise the Commercial interest of this Country. . . . Yet experience may teach us, that the spirit of project and speculation is not always the source of solid advantage. . . . Nothing but highly flourishing manufactures can repay the vast expense of these designs.

Despite this a system of regional canal networks, let alone a national canal system, was slow in developing, and the technology of the canal was anything but symbolic of the industrial age in its continued reliance upon animal traction and in its use of a natural element, water, as basic medium. The flurry of canal promotions in the early 1790s gives a false impression of the chronology of development, for many schemes had long gestation periods, sometimes a decade or more. As consumers of capital and as earth-modelling operations, canals were unfortunately ahead of their time. Capital markets for such large investment projects were as yet ill-formed, while mechanical substitutes for shovel and wheelbarrow remained years ahead (Dyos and Aldcroft, 1969). All this highlighted the critical role of river navigation in early industrial change (Willan, 1936). Rivers remained an essential adjunct of deadwater canals throughout the canal age proper. The early eighteenth century saw many river improvement schemes in the coming industrial heartlands, including the Aire and Calder, the Derbyshire Derwent, the Trent, the Douglas, the Weaver and the Mersey and Irwell

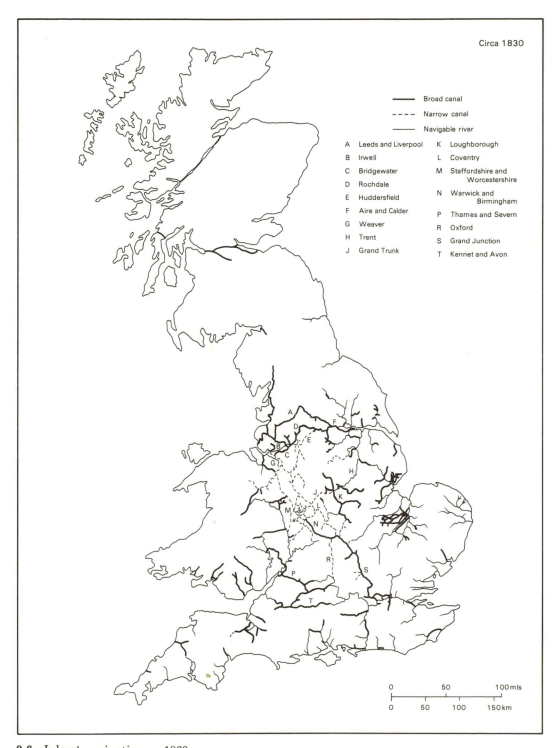

Circa 1830

Broad canal
Narrow canal
Navigable river

A	Leeds and Liverpool	K	Loughborough
B	Irwell	L	Coventry
C	Bridgewater	M	Staffordshire and Worcestershire
D	Rochdale		
E	Huddersfield	N	Warwick and Birmingham
F	Aire and Calder		
G	Weaver	P	Thames and Severn
H	Trent	R	Oxford
J	Grand Trunk	S	Grand Junction
		T	Kennet and Avon

0   50   100 mls
0   50   100   150 km

9.6   Inland navigation, *c.* 1830

(9.6). Most were intended for the distribution of industrial raw materials, coal in particular. Water carriage cost, on average, half that by road. But it was in its release of road transport's capacity constraints that water carriage served above all. By 1750 the country possessed some 1400 miles of navigable river. Canals were the supremely logical extension. By 1830 they had added about 2500 miles (9.6). The most successful canals were typically short, lightly graded cuts which carried heavy mineral traffic: the observation of John Phillips, that every successful inland navigation needed coal at its heels, proved disarmingly prophetic. However, promoters held much wider visions of canalization. From the beginning, schemes for trunk canals were actively pursued. By 1790 the river basins of the Mersey, Trent, Humber, Thames and Severn were linked, albeit circuitously. The early nineteenth century saw three cuts across the Pennines: the Leeds–Liverpool, the Rochdale and the Huddersfield. And by the 1820s, new, more direct canal lines linked London, the Midlands and the northwest. In agricultural districts, coastal feeder canals were seen as means of supplying the population with cheap domestic coal, of facilitating the flow of produce to markets, and of widening the application of farm inputs like lime and marl. Relatively few trunk canals met with financial success and some like the Thames and Severn proved operationally unsound. The agricultural canals fared much worse and some quickly vanished to become beds of railway lines. Even in 1825 there were thirty-seven canal undertakings which barely met their expenses, a far cry from the wildly lucrative Loughborough Navigation, which in the late 1820s yielded an average dividend of 154 per cent.

The key function of canals lay in servicing a quantum leap in the scale and efficiency of regional economic systems. The textbook case in which the building of the St Helens Canal overcame an imminent fuel crisis in the Mersey salt refineries was repeated time and again. As Map 9.7 demonstrably shows, canal traffic was fundamentally short-haul, acting to bind discrete sub-regional resource fields, production sites and markets into a complex, expanding system of symmetric and triangular reciprocities. On the north-western canals *circa* 1790, coal was the most common and vital ingredient of trade in volumetric terms (Langton, 1983). In value it would have been rather less significant; and possibly even less so in subsequent decades as such manufactured items as West Riding cloths poured over the trans-Pennine canals for export from Liverpool. To this extent canals functioned as integrators on a broader geographical scale, but more as export sumps than as transgressors of national economic federalism.

Of the mineral canals, few would have performed as successfully without the intricate systems of feeder roads, waggonways and tramways which lined their routes. It was impossible for canals to serve every pit shaft or adit. They functioned instead as final conduits for the spoils of an essentially punctiform extractive operation. Map 9.9 depicts the feeder tramroads built to serve the Neath and Swansea canals between the 1790s and the 1850s. Most were short-linkage, a function of the characteristic ridge-and-valley topography and the predominant disposition of coal outcrops along the valley sides. Under the Neath and Swansea Canal Acts, mine owners could be granted powers of compulsory land purchase to build tramroads linking their workings with the canals. The lines could be no more than 8 miles distant and had to be available for public use.

Canals were obviously most efficient in lowland terrain. The crossing of watersheds involved serious penalties for efficiency in terms of lockage and water supply, not to mention the escalation in capital costs, while in areas like the northeast the physiography precluded canalization altogether. On Tyneside (9.8) gravity-worked

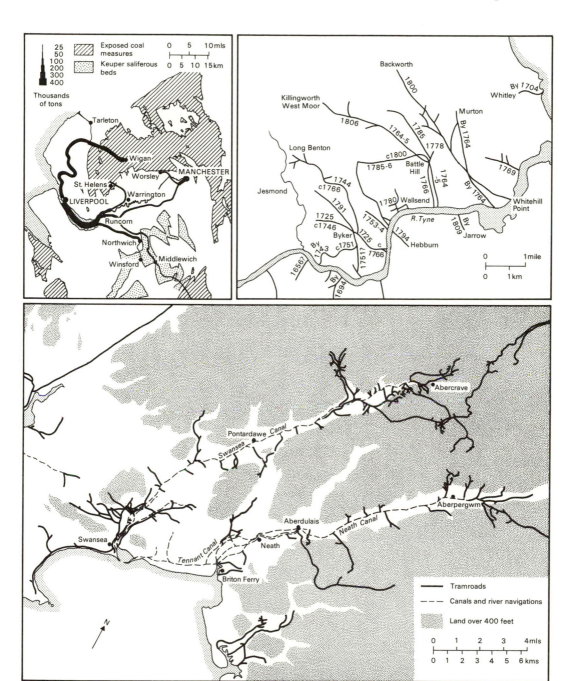

9.7–9 Waterways, waggonways and tramroads, late eighteenth and early nineteenth centuries inclines, wooden waggonways and tramroads provided the means of carrying coal to manufacturing sites and to waiting colliers in the incised valley of the Tyne river (Lewis, 1970). Compared to the valley systems of south Wales, the Tyneside network covered a more extensive area and had a longer history. Here was to be found the seed-bed of the modern railway.

In 1875, Samuel Smiles expressed his views on the dominance of the railway over inland transport in his own particular way:

> The system of British railways, whether considered in point of utility or in respect of the gigantic character and extent of the works involved in their construction, must be regarded as the most magnificent public enterprise yet accomplished in this country – far surpassing all that has been achieved by any Government, or by the combined efforts of society in any former age.

By 1900 railways combined a wide and often dense spatial coverage (9.13) with an enormous range of passenger and freight carriage facilities (Perkin, 1970). Coastal shippers represented their only real competitors, preventing what would otherwise have been a full-scale inland transport monopoly. Railways also broke new ground as industrial and business enterprises. The application of steam power to traction generated large new construction and engineering trades which, in turn, boosted production demands in industries that supplied them (Hawke, 1970). The dozen or so companies which emerged to control and operate the bulk of British railways required vast accumulations of capital. By 1912 they had raised almost £1300m, overshadowing the rates of capital formation of leading industries like coal, iron and steel, and cotton (Gourvish, 1980). As management units, some of the larger companies anticipated the industrial corporations of the twentieth century.

Most early railway projects sprang from the inadequacies of existing means of transport. But when the profitability of these first operations became widely known the railway was rapidly perceived as a focus for speculative investment. Railway mania peaked in the mid-1840s and was largely responsible for the dramatic leap in the extent of the railway network from 1840 to 1852 (9.10–11). Most 'mania' lines had some economic or geographic rationale, but many lines were constructed for which little real justification existed other than to placate hungry bands of investors. By 1860 the basic railway network as we know it today was complete, but construction actually continued for another fifty years (Pollins, 1971). By 1914 mainland Britain had a network of a little over 20,000 miles. Railways had penetrated deep into rural Wales and the Scottish Highlands. The more populous parts of the country were criss-crossed with branch and connecting lines. And even as late as the 1890s a new main line was being laid from the Midlands into London. Some of the network additions made economic and geographical sense, but many did not – the outcome of intermittent continuing speculation and of inter-company competition.

Of the various social impacts of the railway none was more impressive than the startling reduction in the frictional effect of distance on movement. Maps 9.14–16 indicate the scale of the nationwide transformation from the early nineteenth to the early twentieth century. Map 9.17 shows how improvements in Anglo-Scottish rail schedules between 1870 and 1900 effectively shrank the size of the national space by between a quarter and a third. The effect was less dramatic on most other route alignments but the general image became widely appreciated by the business community and, later in time, by the travelling public at large.

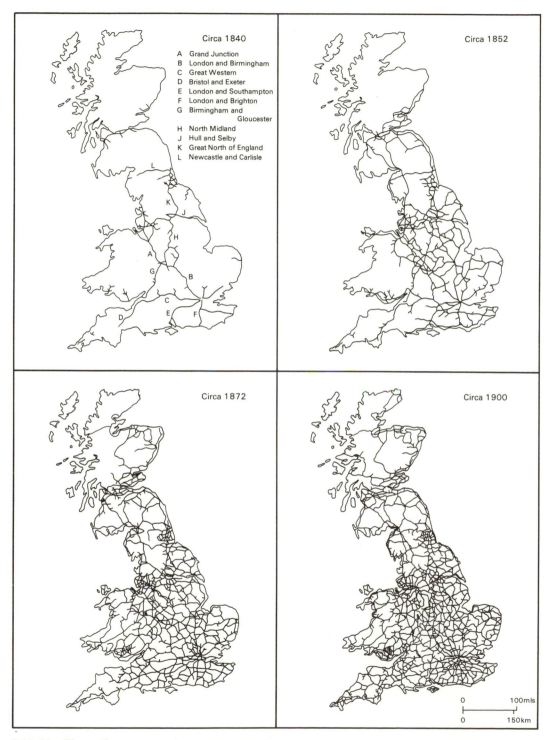

Circa 1840

A   Grand Junction
B   London and Birmingham
C   Great Western
D   Bristol and Exeter
E   London and Southampton
F   London and Brighton
G   Birmingham and
                    Gloucester
H   North Midland
J   Hull and Selby
K   Great North of England
L   Newcastle and Carlisle

Circa 1852

Circa 1872

Circa 1900

0   100mls

0   150km

9.10–13   The railway network, *c.* 1840–*c.* 1900

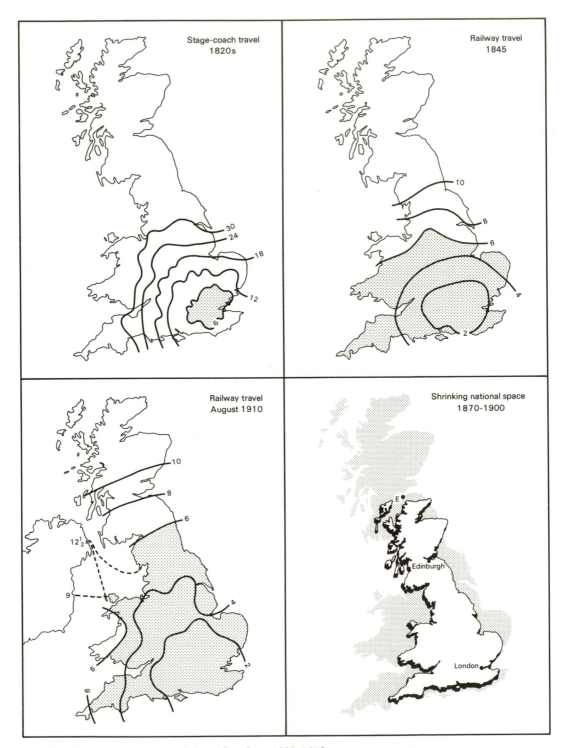

9.14–17 Travel times (hours) from London, 1820–1910

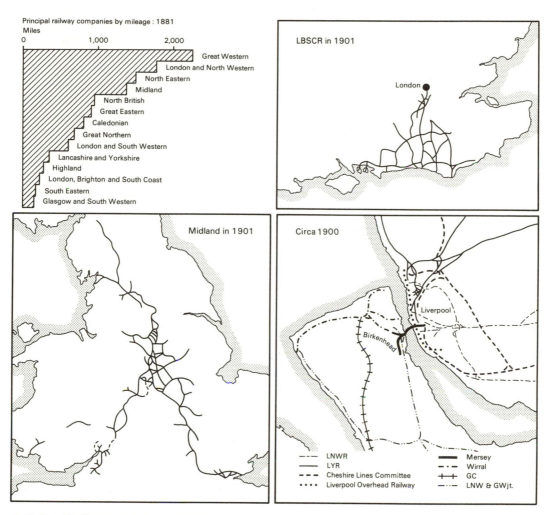

Principal railway companies by mileage : 1881
Miles

0   1,000   2,000

Great Western
London and North Western
North Eastern
Midland
North British
Great Eastern
Caledonian
Great Northern
London and South Western
Lancashire and Yorkshire
Highland
London, Brighton and South Coast
South Eastern
Glasgow and South Western

LBSCR in 1901

London

Midland in 1901

Circa 1900

Liverpool

Birkenhead

LNWR
LYR
Cheshire Lines Committee
Liverpool Overhead Railway

Mersey
Wirral
GC
LNW & GWjt.

9.18–21   Railway company systems

Although the bulk of the railway system was operated by a dozen or so companies, there was great variation in the length of company lines (9.18) and in the geographical structure of companies (9.19–20) (Carter, 1959). The Midland system extended in elongate fashion from London to Carlisle with a spreading web of lines in the area which provided its name and a singular linkage from Birmingham to the Severn estuary. The system of the London, Brighton and South Coast, by contrast, described a highly discrete spatial form to give something near to a regional monopoly. The geographical fragmentation of railway ownership had important economic effects. Most companies found it in their best interests to structure freight rates so as to encourage the trade of the regions they served. The result was often to favour short-distance freight flows. And where companies exhibited discrete territorial identity this acted to cement existing resource grouping, almost forming a policy of regional development. Ownership fragmentation was also manifest in the railway facilities of cities. In the search for passenger traffic especially, companies vied with one another for access such that large cities like Liverpool (9.21) displayed a complex web of through and terminating lines (Patmore, 1961).

Given the highly variable nature of the country's resource and population geography, there were very evident differences in the density of traffic over the separate companies of the railway system (9.22), differences that were often accentuated by the diverse characteristics of the companies themselves with, for example, their varying combinations of single and double trackage (Ross, 1904). Freight traffic was predictably heaviest on those companies serving major industrial districts: for instance the Midland, the Lancashire and Yorkshire, the North-Eastern and the LNWR. The heaviest passenger carriers were mainly found in the southeast of England. By 1900 the LSWR, the LBSCR and the SECR each had extensive commuter traffic. All three companies also offered major passenger gateways to the continent. In summer, meanwhile, their passenger earnings were boosted by the enormous holiday and excursion traffic to the coasts. A few companies carried an extensive passenger as well as freight traffic: for example the Lancashire and Yorkshire servicing the travel needs of the clerical and professional groups which grew in step with the rise of Liverpool, Manchester and Leeds as business and commercial metropolises. The lowest traffic densities were predictably found in the areas of sparse population and low economic activity. On the Highland lines, freight traffic was so small that the disarming critic would no doubt have claimed that trains were more common than traffic.

The backbone of railway freight traffic throughout the later nineteenth century was coal. And because coal was an important article of domestic consumption and also an essential fuel of railway operation itself, it displayed a rather greater average distance distribution than might have been supposed. Nowhere is this more clearly apparent than in the supply of coal to London (9.23). In the 1840s coal came in almost exclusively by sea. By 1855 roughly a quarter arrived by rail and by 1870 over a half. The Yorkshire, Derby and Nottinghamshire coalfields were the primary sources, giving birth to an impressive new commodity flow between the north Midlands and the capital (Hawke, 1970).

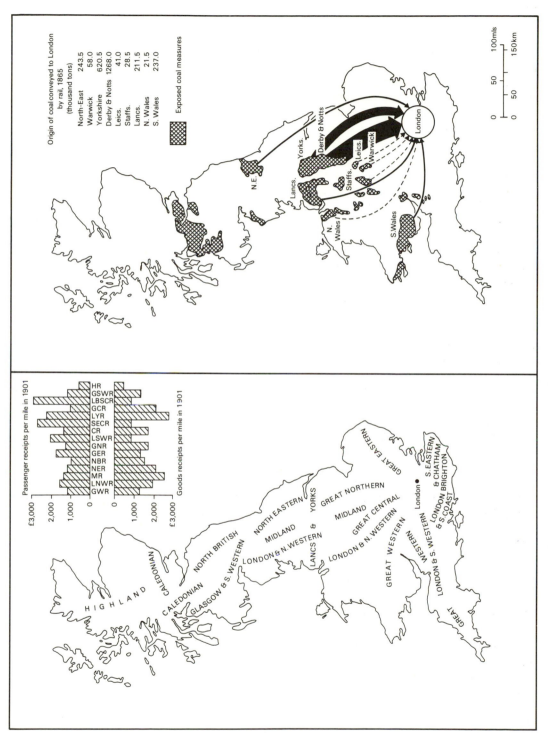

Origin of coal conveyed to London
by rail, 1865
(thousand tons)

North-East	243.5
Warwick	58.0
Yorkshire	620.5
Derby & Notts	1268.0
Leics.	41.0
Staffs.	28.5
Lancs.	211.5
N. Wales	21.5
S. Wales	237.0

Exposed coal measures

N.E.

Yorks.

Lancs.

Derby & Notts

Leics.

Staffs.

Warwick

London

N. Wales

S. Wales

100mls

50

0

150km

50

0

Passenger receipts per mile in 1901

£3,000 2,000 1,000 0

HR
GSWR
LBSCR
GCR
LYR
SECR
CR
LSWR
GNR
GER
NBR
NER
MR
LNWR
GWR

Goods receipts per mile in 1901

0 1,000 2,000 £3,000

HIGHLAND

CALEDONIAN

NORTH BRITISH

CALEDONIAN

GLASGOW & S. WESTERN

NORTH EASTERN

NORTH EASTERN

YORKS

LANCS &

LONDON & N. WESTERN

MIDLAND

GREAT NORTHERN

MIDLAND

GREAT CENTRAL

LONDON & N. WESTERN

GREAT WESTERN

GREAT WESTERN

LONDON & S. WESTERN

London

GREAT EASTERN

S. EASTERN
& CHATHAM

LONDON BRIGHTON
& S. COAST

GREAT

9.22–23   Railway company receipts, 1901; rail-borne coal traffic, 1865

10 Sea trade
Gordon Jackson

In 1700, although Britain was on the geographical periphery of Europe, the economy, especially in parts of England, was by no means underdeveloped in contemporary terms. However, trade patterns still had many primitive features. The former dependence of exports on raw materials and semi-finished goods was being changed by the increased overseas earnings of lighter woollens and other textiles which funded a greater importation of European textile, metal and wooden manufactures on which Britain largely depended. Apart from wool, there were no adequate home supplies of the basic resources required by contemporary industry. There was little good-quality iron or wood and insufficient supplies of dyestuffs, mordants and tannins, potash, flax and hemp, and few oils for the clothiers and soapers. Any pretensions to maritime power floated on imported naval stores. Industrialization, which in time reversed the flow of practically every imported manufacture, relied on a reciprocal exchange of raw materials. On the foodstuffs side, wines, fruit and other luxuries had always been imported.

An important feature of the late seventeenth and early eighteenth centuries was therefore the British merchants' search for new opportunities, gradually applying labour and capital to the exploitation of resources and markets in underdeveloped European countries and lands brought under British control (10.17). Factors from Leith, Hull and London pushed British trade in the Baltic and White Seas, whilst British factories in the Far East were already providing the tea, sugar, coffee, silk and spices that produced a rich store of foreign exchange on their re-export to Europe (Davis, 1962a and b). More important, in the long run, was the increase in markets and national factors of production following the opening-up of the West Indies and North America by British emigrants and African slaves producing furs, fish, rice, sugar, rum, tobacco and eventually timber, ships and the all-important cotton wool. This long-term shift in emphasis towards colonial trade was of crucial significance to Britain (Table 10.1). The non-European share of imports rose from around 30 per cent by value in 1700 (mostly from Asia rather than the infant colonies) to around 55 per cent in 1794/6, when the colonies were absorbing two-thirds of total exports (Davis, 1962a and 1979). This was the great era of re-exports, when the West Indies reached the height of their economic importance and protected colonial trade moved Britain from the edge of Europe to a central position as entrepôt between America, Asia and Europe.

Table 10.1 Value of trade by region (percentages)

| | 1794/6 | | 1844/6 | | 1900 | |
	In	*Out*	*In*	*Out*	*In*	*Out*
Total value	£34.3m	£28.7m	£82.0m	£69.2m	£523.1m	£354.4m
N. Europe	30	28	30	34	37	37
S. Europe	15	11	9	15	9	10
N. America	7	28	24	17	31	13
West Indies	25	18	7	6	0	1
S. America	1	0	6	9	5	7
Africa	1	3	4	3	2	6
Asia	21	13	17	15	9	16
Australasia	0	0	3	2	7	8

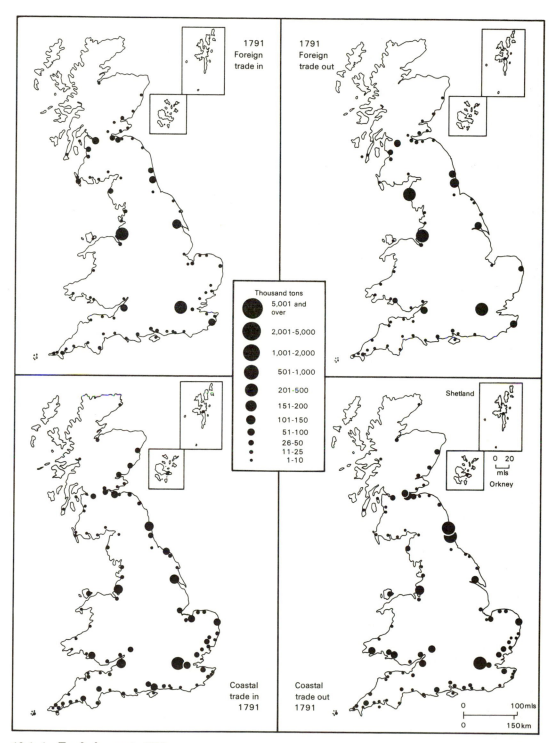

10.1–4   Trade by port, 1791

The rise of oceanic commerce had important influences on regional development within Britain itself which are reflected in the relative performance of the ports (10.1–2). The southeast had long been the most populous and productive region, with many ports serving restricted hinterlands on the coasts between Bristol and Hull and engaging mainly in trade with nearby Europe. While they continued or even expanded their traditional role, they were largely irrelevant to the commercial and industrial revolutions. Despite pier and harbour improvements in the eighteenth century, few made much progress between 1791 and 1841 (10.5–6) (Swann, 1968). London, as the chief population and production centre, had always been the hub of British commerce and was undoubtedly the major instigator and beneficiary of colonial trade, retaining through the East India Company a monopoly of Indian trade till 1813 and Chinese till 1833. The relative position of the capital did decline as geographical advantage enabled Glasgow, Liverpool and Bristol to capture a substantial share of transatlantic commerce, and in the case of the last two to engage in the slave trade (Hyde, 1971; Minchinton, 1957). They accumulated great wealth which was further invested in the processing of colonial materials and the manufacture of goods for colonial markets. Their 'stores' in Virginia for instance secured to Glasgow's merchants the lion's share of the tobacco trade (Devine, 1975). The hinterlands of both Glasgow and Liverpool were transformed through the working-up of imported cotton for sale in colonial and other markets. On a lesser scale many ports in the southwest built up a great interest in the Newfoundland fishery in order to supply dried fish to their traditional markets in southern Europe. A word of caution is necessary regarding the major western ports. The large trade flows were *not* with colonies as might be supposed, but with Ireland, a most important connection which contributed 42 per cent of 'foreign' tonnage entering Glasgow in 1791, 83 per cent at Whitehaven and 33 per cent at Liverpool. Whitehaven had third place as exporter because 98 per cent of its tonnage was carrying coal to Ireland.

The novelty of exotic new trades must not obscure the continuing importance of mundane European trade in which the eastern ports had a natural advantage. As on the west coast, the bulk of trade was shared by the handful of ports on the great estuaries. Raw materials were transformed in the coal-bearing districts of central Scotland fed from Bo'ness and Grangemouth through the Forth and Clyde Canal (1755). During the eighteenth century, the great industrial belt of England from Bradford down to Birmingham was progressively linked to the port of Hull by the improved rivers Ouse, Aire, Calder, Don and Trent and eventually by canals (10.19, inset) (Jackson, 1972).

Apart from the northern coal ports, with their very specialized function, the remaining east coast ports played little direct part in industrialization. Yarmouth and lesser places such as Stockton, Whitby, Bridlington, Wisbech and King's Lynn were active as regional trading centres, as were also several of the isolated ports serving specific hinterlands on the east coast of Scotland, especially Leith, the port for Edinburgh (Lenman, 1975; Wren, 1976). Dover was significant because of the Channel packets.

London continued to dominate foreign trade well into the nineteenth century partly because of a national and international entrepôt function as a break-bulk and/or transhipment point for foreign and colonial goods, while the capital's demand for foodstuffs, fuel, building materials and exportable manufactures attracted a huge coastal traffic in return (10.3–4). In the other major ports the coastal trade was the most important by volume, since almost all long-distance 'internal' traffic was organized by

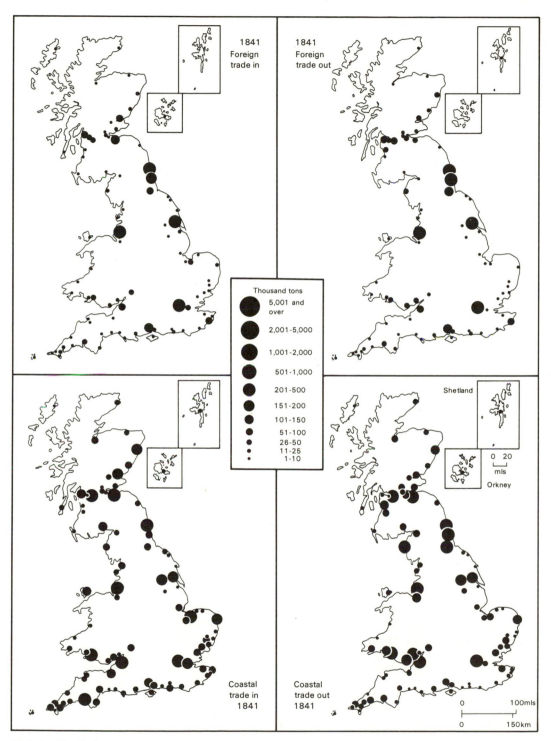

10.5–8   Trade by port, 1841

way of the shortest route to the coast (Bagwell, 1983). A large number of places were involved in a substantial way, especially in the south and southeast where minerals, fish or agricultural produce were shipped and coal and foreign goods received in return. There was no need for them to trade foreign or even between east and west coasts. In any case, few had the merchants or harbour facilities to sustain such trade. Thus, the true value of a port to its hinterland's economy is best indicated by aggregating its foreign and coastal trades.

The frenetic activity of the late eighteenth century appeared at first sight to have come to an abrupt halt after the Napoleonic War. Import and export values grew by only 140 per cent between 1794/6 and 1844/6, a negligible performance compared with the preceding and succeeding half centuries (Davis, 1979), justifying the Jeremiahs who thought Britain's trade was in the doldrums (Table 10.1). This was chiefly a result of falling raw material prices and economies of scale in the machine manufacture of cottons. Exports of cottons rose 13 per cent in value and 337 per cent in volume between 1814/16 and 1844/6. Europe's share of domestic exports doubled, whilst that of North America halved. North America's share of imports trebled but West Indies trade collapsed. The value of re-exports grew only 57 per cent in the fifty years.

It was the volume of trade which filled ships and ports. The contribution of port improvements to declining real prices may never be quantified, but they enabled Britain to handle 4.5 million tons of shipping in 1841 compared with only 1.5 million in 1791. As indicated above, this was heavily concentrated on a tiny group of rapidly expanding ports (Jackson, 1983a). In 1791, London and Liverpool received 54 per cent of the tonnage entering England. By 1841, this had risen to 58 per cent with Liverpool achieving a quarter of the total as London slipped back. The five leading English ports received 74 per cent in 1791 and 79 per cent in 1841. In Scotland, Glasgow's ports had nearly a quarter of the import tonnage in 1791 and 39 per cent in 1841, while Glasgow, Aberdeen, Dundee, Leith and Bo'ness had 55 per cent of the total in 1791 and 86 per cent in 1841. Of English exports, 29 per cent were now handled by eight coal ports and 91 per cent by these and the five leading ports. In Scotland, 81 per cent of entries and 68 per cent of clearances were handled by the leading five.

Such a heavy concentration of activity, augmented by an even larger coastal trade, placed an immense strain on the major ports. As ships got bigger and more numerous, congestion seriously hampered eighteenth-century trade, and periodically thereafter. Berthing in shallow tidal waters was not feasible for large vessels, whilst the greater value of cargoes made security a matter of urgent reform. Some places, such as Newcastle, Sunderland and Whitehaven, coped by building piers and improving harbours, and experimented with gravity feed mechanisms to speed up loading. The best solution was the wet dock introduced for commercial use at Liverpool in 1715. After 1770 this was slowly copied elsewhere as demand exceeded capacity. In some places, local interests such as landowners hoping to improve property values tried to create trade by improving port facilities (10.16). By 1830, dock and harbour developments had taken place on a grand scale at Leith, Hull, Goole, London, Bristol and Liverpool (Bird, 1963; Jackson, 1983b; Pudney, 1975). Rather less grand investments included the coal ports of Kirkcaldy, Newcastle, Sunderland, Middlesbrough, Newport and Cardiff and the new transatlantic packet port of Southampton on the south coast (Daunton, 1977; Temple Patterson, 1971). Where practical, many eighteenth-century authorities put their trust in harbour improvements. There were several large schemes, notably on the Clyde and Tyne in the nineteenth century (Riddell, 1979). Map 10.16 shows a number of thriving

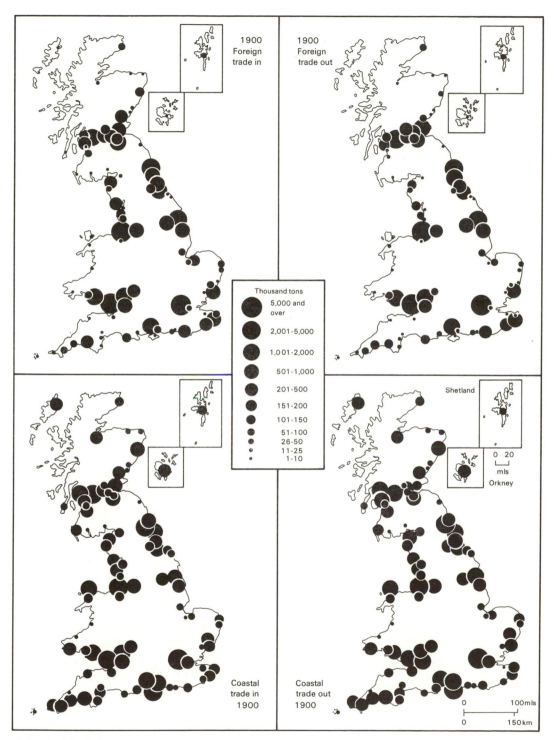

1900
Foreign
trade in

1900
Foreign
trade out

Thousand tons

5,000 and
over

2,001–5,000

1,001–2,000

501–1,000

201–500

151–200

101–150

51–100
26–50
11–25
1–10

Coastal
trade in
1900

Shetland

0   20
mls

Orkney

Coastal
trade out
1900

0   100mls

0   150km

10.9–12   Trade by port, 1900

places of various sizes which constructed harbours outside the existing legal port system. Some became ports in their own right, and others remained adjuncts to existing places. The ambitious attempts of Grimsby and Lancaster to create docks rivalling Hull and Liverpool were a failure (Jackson, 1971). Many successful places shipped coal from Cumberland and the northeast or minerals from the southwest (Beckett, 1981). Others like Grangemouth, Goole and Gloucester were associated with canals (Porteous, 1977). Gainsborough on the Trent enjoyed a brief life as a river port, 1841–81.

The technical and operational limitations on the size of vessels meant that the demands of advancing trade could initially be met by expanding existing harbour or 'first generation' type docks. By mid-century this was no longer feasible. Engineers eager to move coal perfected the railway. Others, eager to ship provisions and passengers coastwise, perfected the steamboat. The construction of the railway network began slowly (9.10–13). For two decades river steamers were accommodated at piers. But, in the 1840s, railways invaded the ports with their voracious demand for landspace, and foreign steamers demanded docks more suited to their increasing size. Specialist steamship docks paved the way for a massive expansion of traffic in the second half of the century. Coasters were generally consigned to inferior harbours and out-of-date docks.

The impact of railways was not confined to existing major ports. They could create and expand ports, such as Methil, Seaham, Hartlepool and Cardiff for coal, and Grimsby and Birkenhead for general trade. Lesser railway ports included Boston, Lowestoft, Harwich, Ardrossan and Troon (10.16). The determining factor in growth was not railways but the demand from shipping. Very few non-coal ports enjoyed great success because the established shipping lines would not use them. The coastal trade was universally aided by trains and steamers and had grown considerably by 1841. The ports of central Scotland, northern England and Wales predominantly engaged in shipping coal and the others in receiving it.

British trade changed noticeably in volume, direction and composition during the second half of the nineteenth century (Saul, 1960). There was a spectacular increase in the size, range and economy of screw steamers, backed by railways and hydraulic power. This facilitated a rise in shipping entering and clearing from 6 million tons in 1845 to 49 million in 1900. Kenwood (1965) has calculated that over £100 million was invested in port facilities between 1851 and 1900, in Liverpool, London and elsewhere.

Table 10.2   Tonnage of shipping by region (percentages)

Region	1790 In	1790 Out	1845 In	1845 Out	1900 In	1900 Out
Total tonnage*	1.43m	1.09m	6.05m	6.03m	49.22m	49.3m
N. Europe	65	56	47	46	58	51
S. Europe	9	9	6	9	11	14
N. America	10	13	29	27	19	16
West Indies	10	12	5	6	0	1
S. America	0	0	3	3	4	5
Africa	1	3	6	3	3	6
Asia	2	2	5	7	3	4
Australasia	0	0	0	0	2	2

Note: *Approximate figures because of changes in calculations over time.

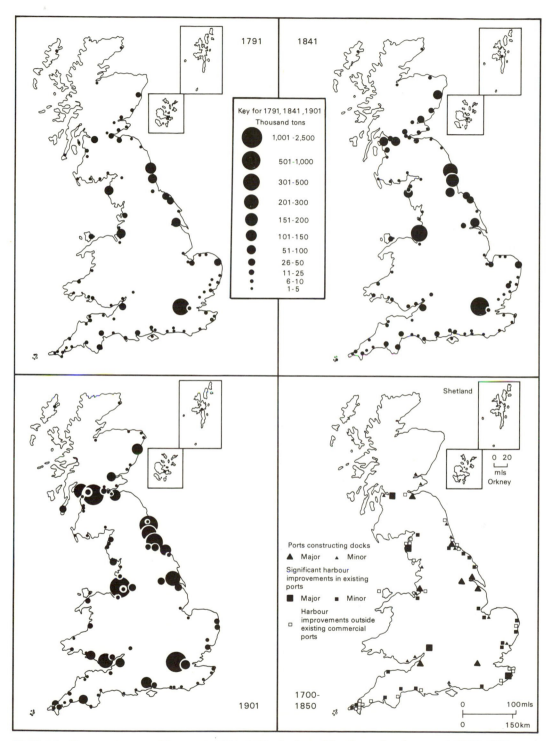

10.13–16 Tonnages owned by port, 1791–1901; dock and harbour development, 1700–1850

The best-known of the steam-orientated developments was the rise of long-distance passenger and goods traffic with the temperate zones already settled by Europeans and with the tropical regions being annexed by imperial powers. Over £3000 million was invested directly or indirectly in a spirited search for materials, foodstuffs and markets. Britain's industry would eventually draw upon Egyptian cotton, Australasian wool, Indian jute, Malayan tin, Latin American nitrates, African copper, American petroleum and Amazonian and Congoan rubber (10.18) (Yeats, 1871 and 1890). More important at the time was the expansion of international food production, with wheat from America, Russia and India, meat from the Plate and Australasia, lard, bacon and cheese from America and tea from India and Ceylon being exchanged for the whole range of British industrial products. The bulk of shipping was not involved in opening up this 'third' world. Asia only sent a fifth of the tonnage of the USA to Britain in 1900. Africa was the same and South America only fractionally more. Their percentage of the export tonnage was slightly higher (Table 10.2).

These new trades did nothing to lessen the importance of northern Europe, which continued to supply the old commodities, supplemented by the 'breakfast trades', bacon, eggs, butter and sugar, and by the new manufactures, 'made in Germany'. By 1900, northern Europe, with 58 per cent of import and 51 per cent of export tonnages, had almost regained the position held in 1790.

This European leadership was not confined to volume. Northern Europe's 37 per cent of both import and export values was higher than the earlier benchmarks (Table 10.1). Imports from North America had risen to 31 per cent but domestic exports to that region had fallen drastically. The empire was not brought in to fill the gap. Exports to Asia, including India, and Australasia had risen only slightly since 1844/6 to 19 per cent and 9 per cent respectively. In 1900, the empire was fulfilling some of the claims of the imperialists so far as the value of trade was concerned, but was still of no great consequence as an employer of shipping and port facilities. Only 2.3 per cent of shipping went to the African colonies and most of that to the Cape.

The continued concentration on Europe was based upon an exchange of raw materials for British manufactures. The connection was consolidated by the inexorable growth of coal exports which transformed the shipping and shipbuilding industries – 1.6m tons in 1840, 7.1m in 1860, 17.9m in 1880, 44.1m in 1900 and 73.4m in 1913. 'King Coal' dominated trade after 1880, filling half the ships cleared at the end of the century. In 1900, Cardiff had the largest export tonnage in the country and with the specialist coal ports, Blythe, Newcastle, Sunderland, Swansea and Newport, shared 46 per cent of export tonnage. The trade stimulated a plethora of supporting railways and gigantic docks, as at Methil and Immingham, which never fully served their intended purpose because of the First World War. Although trade was growing everywhere, it was more concentrated than ever on a minority of places, including the coal ports.

The coastal trade (10.11–12) was now largely a coal trade, much of it simply foreign-going shipping moving in search of cargo or bunkers. It is essential to an understanding of the collapse of British trade, ports and shipbuilding after 1921 to appreciate the enormous contribution of this single low-value and ultimately precarious commodity.

Although London, Liverpool and Glasgow continued to dominate oceanic and imperial trade, the period 1850–1914 witnessed the emergence of many middling ports owing relative success to industrializing hinterlands or better communications. The situation on the Humber was a good example of such changes (10.19). A hinterland

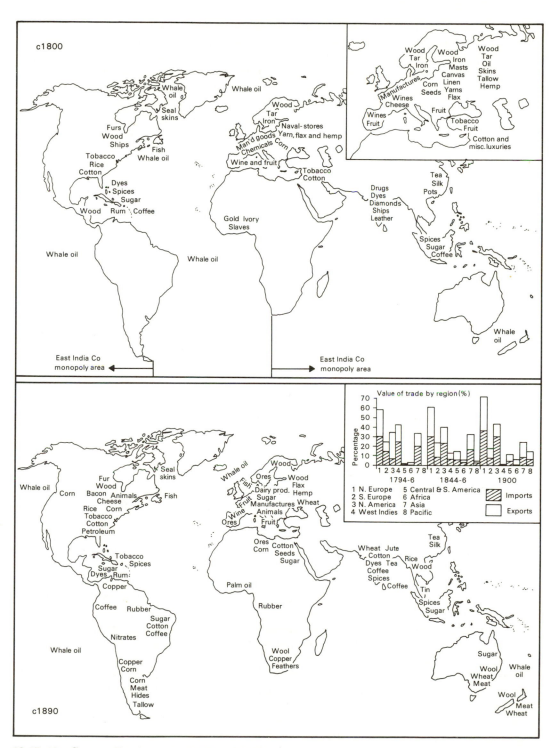

10.17–18   Commodity imports, 1800 and 1890

dependent on water communications with Hull for its eastwards trade was invaded by rival railways determined to syphon off traffic, especially Lancashire manufactures, through the Lancashire and Yorkshire Railway's Goole and the Great Central's Grimsby, while the NER's attempt to divert traffic from Hull to Hartlepool led to the building of the Hull and Barnsley Railway to secure Hull's coal supply (Duckham, 1967; Dow, 1959; Tomlinson, 1915; Hoole, 1972). The expansion of the Yorkshire, Derby and Nottinghamshire coalfield was accomplished with an exceedingly complex railway penetration that further strengthened the two minor ports' challenge to Hull and greatly eased the flow of all sorts of goods into and out of central England (Newham 1913).

This period produced other new and growing ports (10.9–10). Manchester was created to break Liverpool's monopoly of Lancashire's westward trade (Farnie, 1980). Southampton and for a time Plymouth offered transatlantic passengers a swifter journey to London. Harwich provided a faster route to northern Europe, whilst Folkestone and Weymouth served the southern route. Middlesbrough and Barrow were the two chief new industrial ports.

Trade was more complex than this commentary implies and so too was the shipping which carried it. Tiny wooden sailing coasters of 50 to 100 tons were owned everywhere in the eighteenth century and were little different from the short-haul vessels crossing the Channel. Nearly every port had its complement of fishing boats. Sometimes, as at Yarmouth and the Scottish herring ports, these were quite large. For the eighteenth-century Baltic voyages, the 300-tonner was more appropriate and for the East India trade 1000 tons. Like the long-distance trade itself, ownership of these ships was concentrated and built up the tonnage owned by the major ports. Most medium-sized vessels were involved in the coastal and Irish coal trade and were registered in Newcastle, Sunderland and Whitehaven, or in Whitby and Scarborough which specialized in ownership because of their status as harbours of refuge, where the ships were laid up for the winter. Wooden ships could not get much larger for technical reasons, but the faster paddle steamers introduced in the 1820s worked tonnage much harder, so that the growth in registered tonnage between 1791 and 1841 does not indicate fully the increased potential for carrying the larger trade of that period.

By 1901 a number of smaller places had ceased to own ships as inland communications improved and ownership of coastal colliers was concentrated in coal ports (10.13–15). A huge increase in shipowning had occurred since 1841 between Newcastle and Whitby, in South Wales and the Scottish coal ports. The perfection of steel screw vessels, the larger ones averaging 3000 to 4000 tons in 1901, and the largest exceeding 20,000, had increased this concentration of ownership. Only deep-water ports with expensive modern docks could take them, and hence an equally restricted group owned them. London, Liverpool and Glasgow were home to the regular passenger and cargo liners that became household names. Between them, they owned 66 per cent of British shipping and 64 per cent of the steamships in 1900. The largest ships tended to carry passengers. It remained true that much of the carriage of bulky materials, food and manufactures in and out of Britain took place in 1000-tonners involved in rapid transit operations between the Continent, the eastern and some leading western ports. With shipowning as with trade flows through the ports, the general trend between 1791 and 1901 was for success to breed success. Most activity remained in narrowly defined channels, diverted periodically by the development of railways.

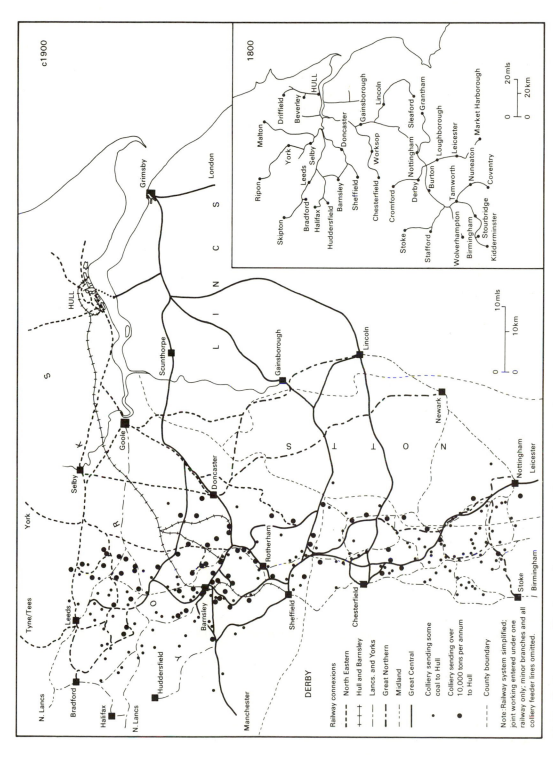

10.19 Railways, collieries and ports in the Trent–Humber basin, c. 1900, and waterways, 1800

11 Textiles

Paul Laxton

Both as a contributor to British exports and as an employer of labour the various branches of textile manufacture remained pre-eminent in the British industrial economy until well into the twentieth century. Until recent decades the steam-driven textile factory has been regarded as an essential touchstone of industrial revolution, consuming in an unparalleled way the lives of men, women and children. So it is easy to be misled by this familiarity with the prominence of textile manufacturing into assuming that we have a readily accessible and comprehensive picture of its size, organization and distribution. The statistical sources are chronologically and geographically patchy. They are irregular in form, inconsistent and sometimes contradictory in content and by no means easy to interpret. Whilst it is not difficult to describe in words, or very crude maps, the broad distribution of, say, cotton manufacture, at least from the 1830s, it is not so simple to quantify the workforce, to distinguish male from female workers, crucial in this industry, to differentiate spinning, weaving and finishing processes, to separate town and country production, to draw comparisons between sets of data based on different sets of geographical areas, or to reconcile the number of employees with the number of installed machines, and installed machines with machines in use. For example, the numbers of workers listed in the Factory Inspectors' reports, the prime source for the distribution of the industry, generally fall far short of the numbers recorded in the contemporary census reports. The ratio between the two numbers varies considerably from place to place. This makes the selection, construction and above all interpretation of maps illustrating the changing geography of textile production very problematical. Any selection can only convey the crudest outline. There is a grave danger that the picture presented reflects the demands of cartographic convenience rather than the problems of economic history.

A dominant characteristic of textile manufacture was its relentless tendency towards geographical specialization and concentration, a process detectable in all branches and through periods of technological change and varying rates of investment. A result of this, at least in woollen, worsted and cotton manufacture, all of which continued to expand until the First World War, was that many places, from substantial towns to small Pennine factory settlements, grew increasingly and dangerously dependent upon textiles for their livelihood. Their vulnerability depended crucially upon both the structure of the local workforce and the particular products and processes upon which they were engaged. Unlike for example the iron and steel industry, the basic regional pattern of production was already firmly established in the early nineteenth century, especially by the time powerloom weaving had become widely adopted in the 1820s. From then on, geographical change, with few exceptions, meant the increasing dominance of core areas of production at the expense of peripheral regions, and the horizontal divergence of different processes, notably spinning and weaving in cotton manufacture. The selection of maps presented here illustrates these tendencies in various ways.

Domestic spinning and weaving penetrated the lowland countryside and remote upland areas, as well as towns in many regions of eighteenth-century Britain and persisted through the 70 to 80 years from the beginning of sustained factory production of yarn to the effective supremacy of factories in all branches by 1850. Maps 11.1–4 depict two such areas where a significant proportion of the male population was engaged in textiles. The Craven district of Yorkshire retained many workers in woollen, worsted

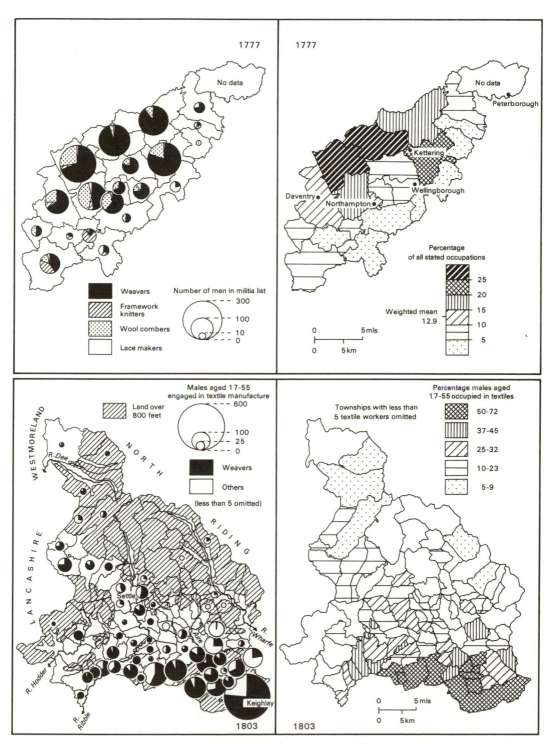

11.1–4 Textiles employment in Northamptonshire, 1777, and Craven, 1803

and cotton goods. Because it was on the periphery of the Lancashire and Yorkshire factory zone it passed through the painful transition from domestic weaving to factory production in the second quarter of the nineteenth century. In contrast, by the 1840s the woollen and hosiery trades of Northamptonshire had completely succumbed to competition from areas further north. In many parts of early modern rural Britain, men and women were, in varying degrees, dependent upon some kind of textile processing, from yarn and cloth to the making of buttons, stockings, nets and ropes. Historians can be wise after the event, but contemporaries in such regions did not see the inexorable rise of the main areas of nineteenth-century production as spelling the inevitable decline of their own textile industries. Nevertheless, such regional shifts, with distressing consequences for uprooted families, were as much a hallmark of the so-called industrial revolution as the rapid and sustained growth of the Victorian mill towns was its post-revolutionary product. If any phase of growth in the textile industry can be called revolutionary it is that for which we lack adequate maps.

The early spinning factories of the Arkwright variety were, for their day, substantial establishments. But many small mills, both for spinning with jennies or small mules, or for fulling, carding and scribbling, were small affairs, distinguished from domestic workshops chiefly by their central source of power, usually a waterwheel. Tracing them through insurance policies, newspaper advertisements and other fragmentary sources is a painstaking task. Map 11.5, taken from the work of David Jenkins, shows the establishment of mills in Yorkshire before 1801, at which point a sharp change of fortune ended a rate of growth rarely if ever repeated in the woollen and worsted industries. Map 11.6 shows the capacity of the cotton spinning industry from 1811 using Samuel Crompton's census, which has been described as 'fairly complete'. Despite its scant access to water power, Manchester was the major manufacturing centre; but within a generation it was to start to decline relative to its satellite mill towns. The number of cotton spindles outside Lancashire and the fact that Crompton did not, to his regret, survey the woollen trades, emphasizes the dangers of studying textile history on a county basis or by type of fibre. It was quite common, especially in the early nineteenth century, for mills to be switched from processing one fibre to another. Our geographical and statistical picture of the early factory industry is very fragmentary.

Not until the accession of Victoria is it possible to present a national summary map of employment in textile factories. Even in Map 11.9, the need to reduce the Factory Inspector's parish statistics to a county level obscures much detail, and the exclusion of Ireland is also unfortunate, as the factory and domestic textile industry there was highly significant for the development of British textiles. Map 11.9 does show how well established was the late Victorian pattern of specialization by 1838. The increased concentration of United Kingdom cotton workers in Lancashire between 1838 and 1900 was largely at the expense of Scotland, Ireland and neighbouring Cheshire rather than the other English counties. The proportion of British woollen workers outside the two leading regions was 29 per cent in 1838 and 28 per cent in 1921. The significant shift was from Gloucester–Somerset–Wiltshire to Yorkshire. More detailed use of the factory returns (Rodgers, 1960; D.M. Smith, 1962; Wilde, 1975; Gregory, 1982) has shown interesting local developments in the 1830s and 1840s, although, as with all analyses of these equivocal returns, such maps should be read with care. The returns disregard domestic textile manufacture, and yet, because they take an operational definition of 'factory' based upon motive power, include many workplaces which employed well

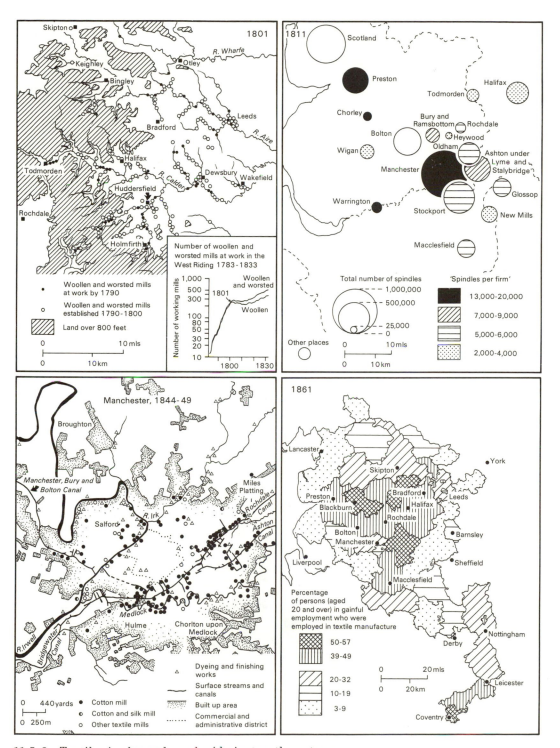

11.5–8   Textiles in the early and mid-nineteenth century

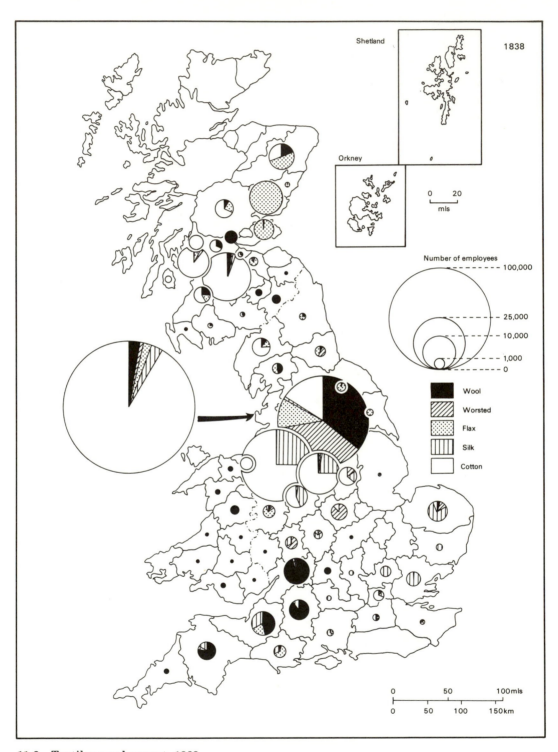

11.9 Textiles employment, 1838

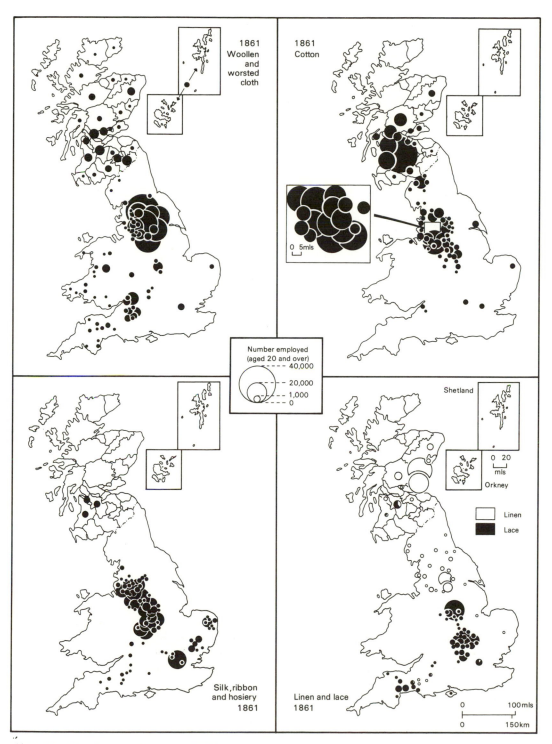

1861
Woollen
and
worsted
cloth

1861
Cotton

0 5mls

Number employed
(aged 20 and over)
- - - - 40,000

- - - 20,000

- - 1,000

- 0

Shetland

0 20
mls

Orkney

☐ Linen
■ Lace

Silk, ribbon
and hosiery
1861

Linen and lace
1861

0 _____ 100mls

0 _____ 150km

11.10–13   Textiles employment, 1861

under 10 people. In 1838, the mean number of workers per British textile factory was 99. It was 142 in cotton and 42 in woollens. Remember also, that these were years of major fluctuations in trade. In the 1838 returns, 14 per cent of the mills in the parish of Manchester were unoccupied, but it is not clear how much this reflected the short-run slump or was part of the long-term shift to the smaller towns. In any event Manchester remained 'cottonopolis', the commercial heart of the industry and a big mill town. Great ranks of mills and dyeworks mustered along the waterways and framed the social geography of the Manchester of Engels and *Mary Barton* (Map 11.7).

A broader view of textile work comes from the employment tables of the census abstract. This takes into account domestic production which was dramatically undermined in some processes by the factory but was stimulated in others until well into the Victorian era. Lack of space means that details of some branches of the industry (see Wilde, 1975 and D.M. Smith, 1963) and of gender differences have been omitted; yet Maps 11.10–13 do reveal striking patterns of concentration in 1861. (The lack of registration district data for workers aged under 20 makes little difference at this scale.) Woollens were most widespread. Flax-based processes were by then largely confined to northern Britain. Lace and silk production were still widely distributed in southern England, where they had been long before the factory era. Map 11.8 distils from these data a most crucial measure, the dependence upon textiles for employment. The registration districts shown constitute the core area of Victorian textile production. They housed 85 per cent of all textile workers in England and Wales and contained several large, densely populated areas with about half of all their workers engaged in textile manufacture and finishing, including hosiery and lace.

Lancashire cotton from 1890 to 1911 provides detail of this local specialization and dependence. The ratio of installed spindles to looms (11.14) defined an arc from the pattern-weaving districts of northeast Lancashire to the coarse-spinning areas around Oldham. Map 11.15 shows, taking account of the general rise in the spindle/loom ratio between 1890 and 1910, that the spinning and weaving districts became still more specialized, while some intermediate places saw greater investment in spindles than average, yet remained, on this measure, which correlates closely with the data on spinners and weavers for county boroughs in the 1911 census, predominantly weaving towns. Two final maps indicate the reliance of workers on these specialisms. Map 11.16 shows both the sex ratio of the workforce and the absolute number of cotton workers: the drift northwards over the preceding century is now clear. In 1911, 62 per cent of the cotton workers were female. It had been 54 per cent in 1835–55. This proportion rose in moving from the northeast to the coalfield in the southwest where the figure was over 85 per cent. Map 11.17 indicates that, even allowing for areas without data, and with the exception of Blackburn, the greatest dependence on cotton work was to be found in the smaller settlements, especially in weaving districts. In some places nearly 90 per cent of women in paid employment worked in the mills. The secular decline of cotton manufacture after 1920 picked off such places with scalpel-like precision as particular products or processes had their markets destroyed. Blackburn, provider of grey cotton cloth for the tropics, was especially vulnerable. In 1936, its unemployment rate reached 27 per cent. The bitter harvest of concentration and specialization, the fruit of a century and a half of slow evolution, was gathered where the crop was ripest.

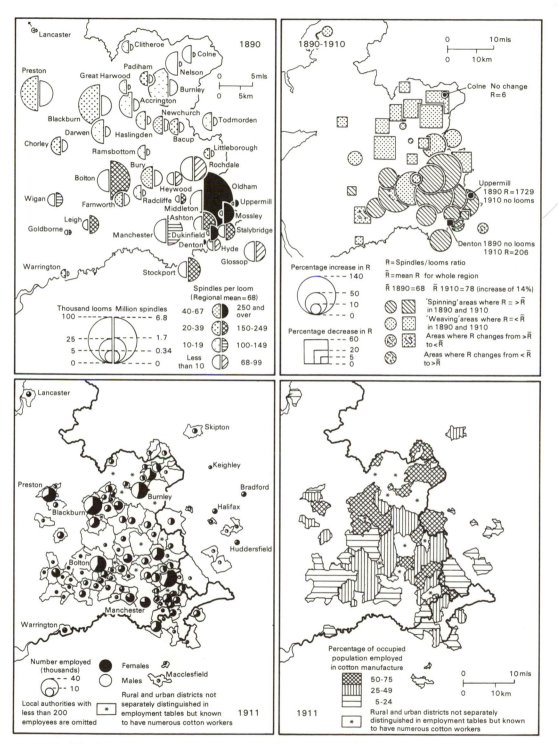

11.14–17 The Lancashire cotton industry, 1890–1911

12 Chemicals

Kenneth Warren

In the twentieth century chemical manufacture has become one of the biggest British industries. Its growth has been among the most rapid and sustained. It operates some of the largest plants and the leading chemicals firm is first in rank among largely home-based private industrial concerns. In marked contrast, during the 160 years to the First World War the chemical industry was much less prominent as compared for instance with coal, iron and steel, shipbuilding or textiles. However, it was laying important foundations for the contemporary industry and for its distribution.

Although many nineteenth-century trades had a wide diversity of final products, most used a limited range of materials and turned out either a finished or an intermediate product which gave them a common basis for measuring output and growth, such as lengths of cotton thread or cloth, displacement of shipping launched or tonnage of puddled bar or ingot steel. In chemicals, on the other hand, materials, processes and intermediate products, as well as final products, were all so diverse that the only possible overall measures of change were value of output or size of workforce. Then as now the chemical industry was a congeries of trades rather than a trade (Hardie and Pratt, 1966). Technically their only common feature was that they employed processes of chemical change to convert raw materials into consumable form. The metal industries in this sense were the biggest chemical trades, but, both because of the type of product and still more the size of the industry, they are conventionally excluded.

From the start of the industrial revolution, the chemical industry employed some materials from distant, often overseas sources of supply – kelp from the wild northern or western shores of the British Isles, barilla from the Mediterranean, potash from North America or the Baltic, sulphur from Sicily. However, then and for long after, it also drew on a wide range of more accessible home resources – alum shales, copperas, pyrites, gypsum, limestone, coal and common salt. Just as a number of British landscapes bear relics of the old chemical processing trades, so also fields and hillsides carry the scars from the working of their raw materials (12.1). The relationships of the chemical trades with other industries were often close. Galvanizing and tin-plating were important throughout the nineteenth century as growing markets for acid, and textiles, soap and glass, provided major outlets for alkali and in the process were a source of a good deal of the capital and entrepreneurship for that industry's early growth. Coal, coke-oven and gas industries were significant suppliers of chemicals, as by-products. With some metal trades there was a two-way relationship. Thus copper smelters provided sulphur as a by-product, while the pyrites burners of the alkali manufacturers brought them in turn into copper production as a profitable sideline. The same sort of relationship is discernible with iron and steel. Availability of iron-rich waste from the pyrites processing operations was in the 1870s the initial locating factor for the pioneer major Scottish steel firm; from 1883 the basic Bessemer process provided basic slag as a new contribution to the array of chemical fertilizers.

Another theme of significance in the growth of the chemical trades to the First World War was the powerful force of technological change and with it the introduction of new materials and a general increase in scale economies, and, for both reasons, the choice of new locations for chemical manufacture. Fewer, bigger locations replaced an earlier, wider scatter of small plants. The oldest technology, that of the pre-industrial or early industrial age, involved either the direct use of natural materials or small-scale and relatively simple conversion processes of organic resources. Thus bleaching was by

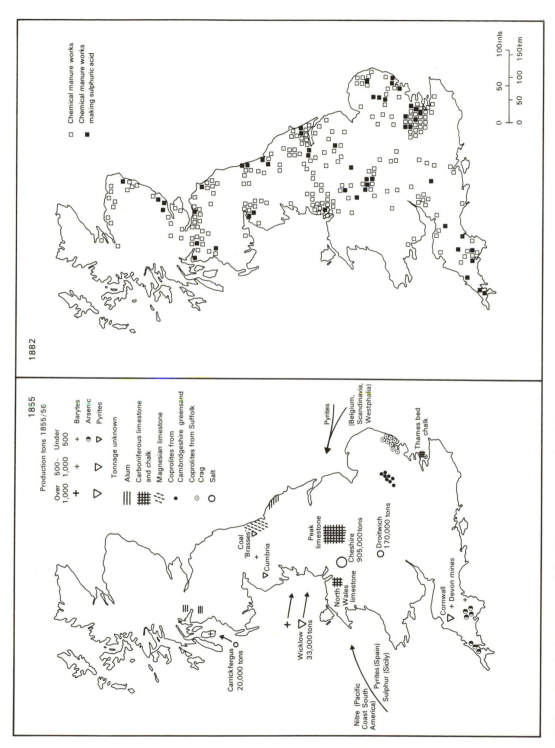

12.1–2 Chemical raw materials, 1855; manure works, 1882

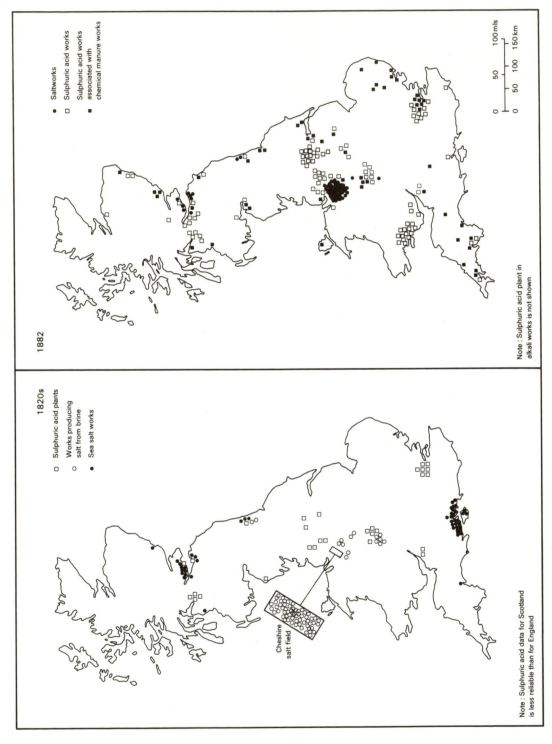

1882

Saltworks

□ Sulphuric acid works

■ Sulphuric acid works associated with chemical manure works

● Works producing salt from brine

○ Sea salt works

1820s

□ Sulphuric acid plants

Cheshire salt field

Note : Sulphuric acid plant in alkali works is not shown

Note : Sulphuric acid data for Scotland is less reliable than for England

12.3–4   Salt and sulphuric acid plants, 1820s and 1882

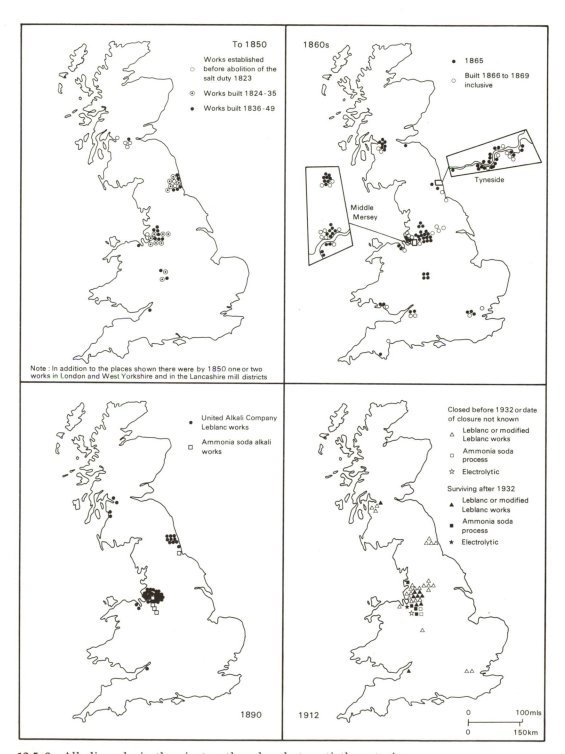

12.5–8 Alkali works in the nineteenth and early twentieth centuries sunlight, and dyestuffs and alkali manufacture involved the use of resources such as madder or indigo in the first case and of kelp, barilla and potash in the second. Farmers employed natural manures. The small-scale demand for vitriol was met by the long-established processes of the distillation of copperas. In the course of the industrial revolution, new, larger-scale and usually mineral-based processes replaced these older ones. Thus, in time, the lead chamber process, first used in Birmingham in 1746, transformed the manufacture of sulphuric acid (12.2–4); Tennant's bleaching powder ousted the bleaching field during the early years of the nineteenth century; following the pioneer work of Liebig and of Lawes, natural manures (12.2) and then guano imports were gradually supplanted by the manufacture of superphosphates made from bones and imported phosphate rock treated with sulphuric acid. A generation later sulphate of ammonia began to come into use, derived from gas works, Scottish iron works, and later still, from by-product coke ovens.

The old alum shale mining trade of Cleveland was killed by Spence's process which opened the coal measures as a source of supply for alum. In the second half of the century synthetic dyes derived from coal tars replaced the natural dyes and, in the process, shifted supply from British to German sources. At the end of the period a 'neo-technic' age began and heralded the chemical triumphs of the twentieth century. Thus from the mid-1870s Solvay's ammonia soda process began a rapid supplanting of the Leblanc process (12.7–8) and in the 1890s electrolytic fixation of atmospheric nitrogen began. Pioneer work on the contact process for sulphuric acid was done in the 1890s and by the early years of the new century artificial silk (rayon) marked the beginning of the chemical-based man-made fibre industries. Before the First World War Haber's process for ammonia production had been made commercially viable and a decade later it was being developed on a large scale in Britain. In the same years in the United States there were occurring the first slight stirrings of the petrochemical industry which was to become so vitally important half a century later.

The newer chemical centres of the late nineteenth and early twentieth century were generally oriented either to the bigger, richer, sources of raw materials, as with alkali to the salt fields of Cheshire, or to markets, as with the artificial fertilizer trades or the dyestuffs industry of Manchester or of Huddersfield, or were near to import or to export points as with a variety of other chemical trades (Warren, 1980). One of the striking features of the geography of the late twentieth-century industry is how much of the nineteenth-century distribution pattern shows through the overlay of another three-quarter century of growth, technical change and business rationalization. The Northwich district at the centre of the old salt field now contains the whole of Britain's heavy alkali industry. Widnes and Runcorn, boom areas of the mid-nineteenth-century trade, remain prominent in sulphuric acid and chlorine manufacture, and, notwithstanding such giant centres as Billingham or Avonmouth or Immingham, many traces of the small, scattered rural chemical manure operations so characteristic of the late nineteenth century are still discernible.

13 **Brewing and distilling**
R.B. Weir

Brewing and distilling have a major place in the history of the industrial revolution. Until the advent of the temperance movement in the 1830s they produced commodities which were widely regarded as essential for human existence, and as a result they accounted for a high proportion of consumer expenditure. Buoyant demand made them attractive to the taxman and a significant part of the public revenue was raised from alcoholic drinks and their raw materials. A growing market also created ideal conditions for technological change and the large-scale breweries and distilleries of urban Britain were amongst the earliest users of steam power. Of even greater importance to contemporaries was the effect of the demand for drink on agriculture. Public policy, through higher taxation on imported drinks and raw materials, sought to ensure that it was British farmers and landlords who benefited from the derived demand for beer and spirits. The repeal of the Corn Laws in 1846 removed the underlying protective justification for such measures but it was not until 1880 that the last of the restrictions which had kept the drink producer linked to the home farmer was removed. All the barley-growing regions, from the advanced farming of East Anglia and East Lothian to the more traditional peasant agriculture of the northeast Highlands of Scotland, experienced the benefits of demand from the drink trades (4.9–12). In their turn the drink trades contributed to the food supply through waste grains for animal feeding. Large-scale distilleries and breweries often maintained dairy herds and fattened pigs and beef cattle. The very seasonality of brewing and distilling with production starting in September or October, after the harvest was in, and being completed in May or June, before the next harvest, symbolized the complementary nature of the drink trades and agriculture.

Consumer tastes were not uniform. Beer was the most popular drink in England. In the late 1820s, consumption per head was six times greater than in Scotland. The English spirit drinker preferred gin to whisky, whilst in Scotland two types of whisky were consumed: grain whisky, made from a mixed mash in which only a small proportion of the barley had been malted, and malt whisky, made from a mash entirely composed of malted barley. These different products, evident by the 1790s, initially owed much to the attempt by distillers to offset malt duty, but the distinction was reinforced after 1830 by the invention of the Coffey or Patent still. Unlike the traditional pot still which was (and is) a batch process, the Coffey still was a continuous process (Weir, 1980). Until 1851, differences in consumer tastes were accentuated by different levels of spirit duty between England, Scotland and Ireland. These features make separate treatment of brewing and distilling desirable.

Map 13.1 shows the distribution of Scottish distilleries by excise collection in 1798/9. From the 1780s the Excise pursued a policy of concentrating output in large-scale distilleries where duty would be easier to collect and prohibiting distilling for private use from small stills. Given that total Scottish spirit consumption was estimated at 3.5 million gallons in 1798/9 and that total licensed output was only 844,017 gallons, the policy clearly failed, a conclusion reinforced by the upsurge in illicit distilling north of the Highland line (Devine, 1975). For peasant farmers spirit was a cash crop of higher value and more easily transported than grain. That it paid the rent explains the reluctance of the landed gentry to curb smuggling. Prosecutions give a partial indication of the regional nature of illicit distilling and the extent of a trade which developed a high degree of specialization between producers and distributors.

Comparison of the number of distilleries and the scale of output shows distinct contrasts between licensed producers in the Highlands and the Lowlands. The four licensed distillers in the Inverness collection, for example, produced less than 50,000 gallons between them whilst the two distilleries in the Edinburgh collection produced nearly 200,000 gallons. These differences can be explained by access to a much larger urban market, greater capital resources, and an excise system which based duty on the cubic capacity of the still and gave distillers an incentive to maximize throughput.

The increasing volume of illicit distilling provoked demands for reform and from 1816 the Excise adopted a policy designed to meet the needs of the small-scale producer with reductions in license and spirit duty. The 1823 Distillery Act, a precocious application of *laissez-faire*, was, as Map 13.2 suggests, extremely successful. Licensed output increased to 9.147 million proof gallons (mpg) by 1833 with 258 distilleries 'entered' for production. For the Excise the most gratifying results were in the regions where illicit distilling had been most in evidence before 1823. As in 1798, however, the same contrast in the scale of production was evident (13.3). Co-operative distilling in which several farmers shared a still was a popular response to agricultural depression but in the long run such enterprises seldom proved viable. Few former illicit distillers succeeded in adapting to a commercial environment. Severe competition, shortage of working capital, poor marketing and, not infrequently, the poor quality of the product produced a sizeable crop of failures.

Excise reform drew in small units of capital but the long-run trend was towards larger units of production. By 1869 there were 123 distilleries in Scotland with a total output of 12.6 mpg but 55 per cent of output came from twelve patent still distilleries. The period 1823–60 thus saw substantial structural change in the Scottish distilling industry, much more so than in England where the maximum number of distilleries was seventeen, the majority located in and around London with important provincial centres in Liverpool and Bristol. Their main activity was making grain spirit which was supplied to rectifiers for conversion into gin after the addition of flavouring materials (juniper berries, coriander seeds, herbs, etc.).

Map 13.4 summarizes the position in 1900 when output reached its peak level. The dominant feature after 1870 was the growing popularity of blended Scotch whisky (mixtures of malt and grain) sold under proprietary names such as Dewar's, Haig's and Teacher's. Blending transferred the final stages of manufacturing and distribution to the Lowlands. As the transfers of spirit between the different parts of the United Kingdom suggest, the English market was of considerable importance for Scottish (and Irish) distillers. Exports were also buoyant though the United Kingdom imported more spirits than it exported.

Brewing was largely spared the trauma of frequent change in excise policy and, as the geographically widespread nature of the industry in the mid-eighteenth century suggests (13.5), the Excise was much more successful in collecting revenue from beer than from whisky. The important position of London also stands out (Mathias, 1959). There a densely packed urban market first enabled brewers to overcome the barrier to economies of scale imposed by high overland transport charges on a cheap and bulky commodity. Cost-cutting assisted the shift in consumer preference from ale to porter, a beer first brewed in 1722. Comparison of maps 13.5 and 13.6 indicates the continuing expansion of the industry in which London's common or wholesale brewers – 'the powerloom' brewers of the industrial revolution – played a leading part. In 1822, the

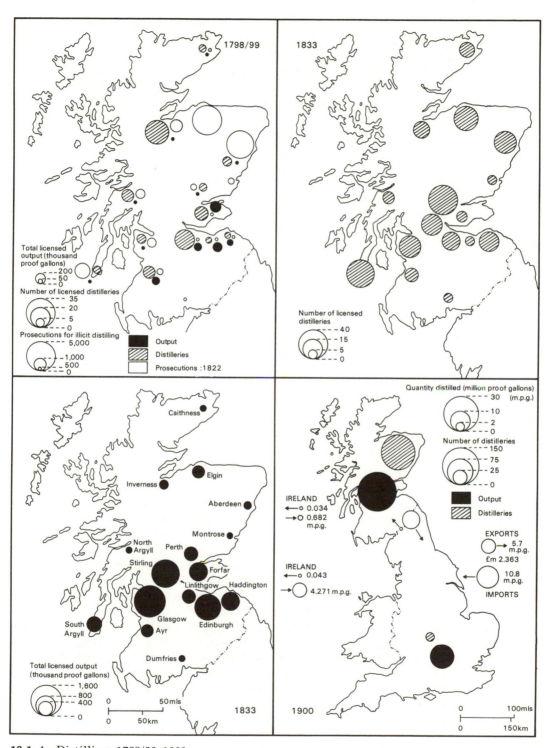

1798/99

Total licensed
output (thousand
proof gallons)
200
50
0
Number of licensed distilleries
35
20
5
0
Prosecutions for illicit distilling
5,000
1,000
500
0

Output
Distilleries
Prosecutions :1822

1833

Number of licensed
distilleries
40
15
5
0

Caithness
Elgin
Inverness
Aberdeen
Montrose
North
Argyll   Perth
Stirling
Forfar
Linlithgow   Haddington
South   Glasgow   Edinburgh
Argyll   Ayr
Dumfries

Total licensed output
(thousand proof gallons)
1,600
800
400
0

0   50mls
0   50km

1833

Quantity distilled (million proof gallons)
30   (m.p.g.)
10
2
0
Number of distilleries
150
75
25
0

Output
Distilleries

IRELAND
←o 0.034
→o 0.682
m.p.g.

IRELAND
←o 0.043
→o 4.271 m.p.g.

EXPORTS
5.7
m.p.g.
£m 2.363
10.8
m.p.g.
IMPORTS

1900

0   100mls
0   150km

13.1–4   Distilling, 1798/99–1900

ninety-eight wholesale brewers in London represented a mere 6 per cent of the total number in England and Wales but they accounted for 43 per cent of the production of strong beer. Outside London structural change was less evident. In Yorkshire, for example, two-thirds of beer output was produced by licensed victuallers (retailers who brewed beer) and it was not until after 1840 that the common brewer began to dominate output (Sigsworth, 1967). Map 13.6 also indicates the minor place of brewing in Scotland relative to England and Wales.

In 1830 the heavy beer duty was terminated and replaced by malt duty, so that Map 13.7, which shows the distribution of malt used in brewing by county in 1850, is a measure of raw material used rather than final product. There was, however, a reasonably close correspondence between malt and output until 1880 when Gladstone 'freed the mash tun' by permitting the use of other raw materials.

Total beer consumption in the United Kingdom rose from 500 million standard gallons (msg) in 1830 to 540 msg by 1850 but within this total there was a shift in consumer tastes away from porter and towards lighter pale ales and bitter, a change which favoured locations where the nature of the water supply was suited to the production of these beers. Burton-upon-Trent and Tadcaster were both examples of such geologically determined locational advantage. They were also, like Alloa and Wrexham, towns whose growth depended upon brewing, a reversal of the more usual situation where the expansion of brewing depended on the growth of towns. The creation of the railway network enabled brewers to cater for a national rather than a local market, and the force of competition during the 1860s and 1870s was reflected in a continuing decrease in the proportion of malt brewed by licensed victuallers and other beer retailers, from 45 per cent in 1830 to 10 per cent by 1890. The number of wholesale brewers also fell. Output was becoming increasingly concentrated in fewer firms, a process aided by the re-introduction of restrictive licensing in 1869 which encouraged the acquisition of licensed property by brewers.

Per capita consumption of beer peaked at 34.4 gallons in 1876 but population growth, and to a lesser extent the expansion of exports, meant that total output rose from 1145 msg in 1876 to 1326 msg in 1899 (Wilson, 1940). As in distilling, this peak was not to be passed again until well after the Second World War. Despite the changing social attitude to drinking brought about by temperance reformers and 'the Brewers' Wars' over tied houses, the closing decade of the nineteenth century was a time of considerable prosperity for the industry. Map 13.8 thus depicts the industry at what was then its greatest extent, sharing its prosperity with supplying industries such as hop growing and malting, though not with British barley growers, as brewers, like distillers, increasingly switched to foreign sources of supply for barley.

Maps 13.1–6 are based on excise collections, i.e. the administrative unit for the collection of spirit and beer duty. The precise boundaries of the collections are not known and varied over time but, used carefully, the collectors' returns provide a good indication of the changing spatial distribution of the distilling and brewing industries. Maps 13.7–8 are based on counties, that is, the collections have been grouped. There was reasonably close correspondence between county boundaries and excise collections except in one important case, Burton-upon-Trent (Derbyshire) whose collection was Lichfield in Staffordshire.

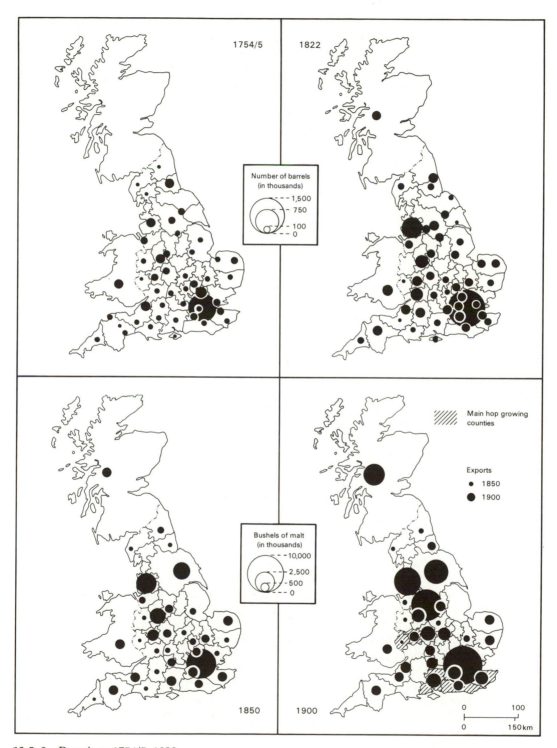

13.5–8 Brewing, 1754/5–1900

14 Leather footwear

P.R. Mounfield

Before the mid-nineteenth century shoemaking was a widely practised handicraft; each community made its own, often using local materials. In the towns, cordwainers' workshops and shoemakers grouped in trade quarters. This division of labour was most advanced in London. By 1650 only London and Northampton manufactured for more than local demand. Before the introduction of effective machinery in the 1850s, wholesale manufacture was superior to traditional bespoke methods, mainly for the fulfilment of large orders for a relatively standardized product, such as was needed by the army. From the mid-seventeenth century onwards, large army orders enabled Northampton master shoemakers to enter wholesale production. Other centres in the county joined the industry throughout the next 150 years (14.4) (Mounfield, 1965 and 1968). Elsewhere, wholesale shoe manufacture was begun by William Horton in Stafford in 1760 and in Norwich by James Smith in 1792.

By 1851 many of these places had warehouses acting as foci for the organization of outworking, but they contained little machinery or power. Demand was increasing, however, at home and abroad and railways were providing access to a national market. Although power needs were relatively modest, machines of various kinds were entering the production process. Wholesale production could begin wherever there was labour and entrepreneurship available. In 1833 James Clark began to make slippers in Street, Somerset; in 1842 R. M. Somervell started *K* Shoes in Kendal, Westmorland; in 1845 the Derham brothers began in Bristol; in 1851 Thomas Crick was making shoes for stock in Leicester; in 1870 J. W. Rothwell and Samuel McLerie started production in two old textile mills at Waterfoot, Rossendale. By the 1880s factory manufacture was firmly established in these and other places. Increasing geographical concentration of footwear production was accompanied by a movement of labour out of workshops (14.1–2). In 1895, 85,571 footwear workers in England and Wales were employed in factories, compared with 37,535 in workshops: by 1905 the number in factories had increased to 102,489, compared with 21,703 in workshops.

In 1911 (14.3) Northamptonshire and Leicestershire contained over half the country's footwear manufacturing industry. Northampton had early contact with the London market and army clothiers and reasonable transport links with the capital, enabling the use of plentiful local supplies of good quality leather and an underemployed labour force. It became known early on as the most important place in the country for men's wholesale hand-sewn work. Leicester's footwear industry was prompted by technological innovation. In 1853 Thomas Crick patented a method of riveting shoe soles to their uppers that was quicker, cheaper and required less skilled labour than even the lowest quality of welt-sewn goods. Others emulated him and from 1853 to 1867 the number of footwear factories in the town rose from four to seventy. Other places in the county such as Anstey (1863), Sileby and Earl Shilton (1870) and Barwell (1877) became involved, too. Women's and children's shoes were Leicestershire's speciality.

In Leicestershire, labour supply and low wage-bills for footwear entrepreneurs were closely associated with the decline of a preceding domestic industry, hosiery. In eastern Northamptonshire, at an earlier date, particularly along the Ise Valley, footwear had replaced woollens, silk and lace. In both counties the initative of particular individuals can be identified at the early stages and, in both, later developments outside the county towns were affected by the routes adopted for railway lines during the nineteenth century.

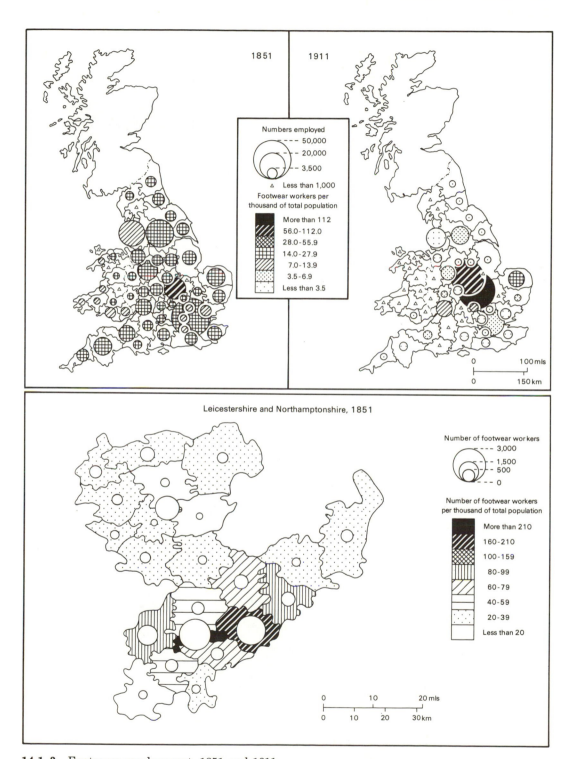

14.1–3   Footwear employment, 1851 and 1911

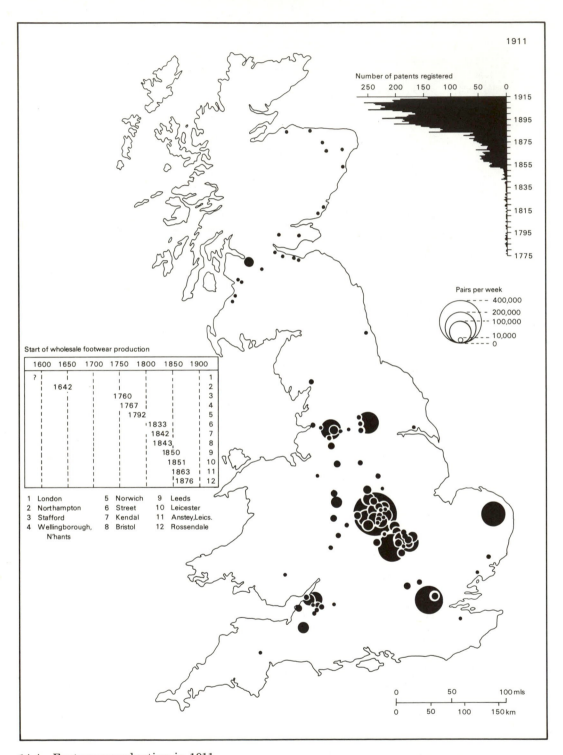

14.4   Footwear production in 1911

15 **Iron and steel**
Philip Riden

The historian of iron and steel is fortunate in that first the trade itself and later the government took a keen interest in the size and structure of the industry, so that statistical data relating to domestic production, as well as imports and exports, are available from the early eighteenth century. The usual basis of output estimates, indeed the only series which can be used for long-run comparisons, was the tonnage of pig iron produced in the blast furnace each year, even though pig was (and is) an intermediate good whose only use is as the raw material for subsequent reworking. In the early eighteenth century about 90 per cent of pig produced in Britain was refined into a tough, malleable, decarburized material known (from its shape) as bar iron. The proportion of pig used for castings gradually rose, as forge technology stagnated while foundry technique improved, until at the beginning of the nineteenth century almost half the output of pig went into castings. Forging was then transformed and by 1830 pig output was distributed in proportions of roughly 70:30 between bar and cast iron, as it had been half a century earlier. After 1856, and more especially 1870, the industry was revolutionized once again as mild steel, essentially an alloy of iron, carbon and small quantities of other elements, became the main end-product of the industry. Paradoxically it was only in 1881, when its decline was already evident, that statistics of puddling forge output began to be collected.

A series of six maps (15.1–6) illustrate the changing regional distribution of pig iron output between about 1720 and the First World War. In the early eighteenth century Britain produced some 20,000–25,000 tons of pig annually, a figure which rose steadily over the period 1660–1750, despite a lack of technological change and, in particular, a continuing failure to substitute cheap mineral fuel for charcoal. About 27 cwt of pig was consumed in the production of a ton of bar iron at this period, so that (allowing for between 5 and 10 per cent of output to be used for castings) bar iron output may have been about 15,000–18,000 tons a year (Riden, 1977). This represented no more than half, and possibly only a third, of Britain's total consumption of bar iron, which was rising considerably faster than domestic output. The remainder was supplied by imports, principally from Sweden, whose primacy, established in the middle decades of the seventeenth century, was increasingly challenged in the eighteenth by Russia. Small amounts of iron continued to be imported from Spain. If anything, the position of the domestic industry in the home market weakened in the first half of the eighteenth century.

Most iron smelting in this period was distributed over a large, loosely bounded region whose unifying characteristic was proximity to the Bristol Channel or the Severn, navigable from its mouth to Welshpool and a river of great importance to the industry. Furnaces were to be found in most counties in a broad arc from Carmarthen to Cheshire, with concentrations in the Forest of Dean and Shropshire, exploiting easily accessible deposits of relatively rich ore. Their dependence on water for power and charcoal for fuel largely determined their exact location within ore-fields and militated against the building of more than one furnace on a single site. Most operated in tandem with one or sometimes two forges, which might in turn, especially in the West Midlands, supply water-powered rolling and slitting mills, which produced 'rod iron' in long thin strips for nailmakers, the most important single group of customers for the industry in this period. The only other important regional concentration in what was, by later standards, a widely dispersed industry, was in the West Riding, Derbyshire and Nottinghamshire, the area served by the Trent, Don and other rivers for which Hull was the outport, in which some 12 per cent of output was produced; the Severn valley and its outliers accounted for about 65 per cent. The remainder came from several minor districts, including north Lancashire and Scotland, where blast furnaces were only now replacing bloomeries, and the Sussex Weald, once the country's most advanced iron-making region, now left with about 10 per cent of output, the bulk of which went into gun founding and other foundry work.

During the second half of the eighteenth century the iron industry was transformed by a series of technological changes from dependence on water power and charcoal into the major customer for coal (in the form of coke) and an important user of steam power for blowing bellows and driving machinery (15.1–2). The industry gradually captured more of its home market, until by 1820 only high-grade Swedish bar iron used in the Sheffield steel industry had still to be imported. By 1800 charcoal had all but ceased to be used in the blast furnace, surviving only in the Forest of Dean and south Wales, mainly to meet the specialized needs of the local tinplate industry. There were lesser survivals in Furness and the Weald. With the adoption of coke, furnaces were now concentrated in those coalfields where iron ore was also found. Above all, this meant the northern rim of the south Wales coalfield of Monmouthshire and Glamorgan, which, with outliers in Breconshire and Carmarthenshire, produced nearly a third of total output. Shropshire, which in the mid-eighteenth century, when coke was first adopted as the fuel in the blast furnace, was the country's most important iron-making region, had lost that position and now accounted for about a fifth of the total, as did south Staffordshire and the Black Country (Hyde, 1977). The West Riding was responsible for some 10 per cent, twice that of its neighbour, Derbyshire, while the iron industry in central lowland Scotland, although much larger than in the eighteenth century, produced just under a tenth. The geography of bar iron production (or puddled iron, as it became known after the adoption of a new coke-fired process in place of the old finery-chafery technique) was much the same as that of smelting: the eighteenth-century pattern of dispersal gave way to one of concentration of all processes on a single site and large numbers of works in each district.

Between 1800 and 1830 the iron industry suffered gigantic convulsions in output with the change from war to peace in 1815 and the first stirrings of railway building in the 1820s, coupled with a legacy of over-capacity from the period of rapid expansion between 1790 and 1810. Forge technique improved during these years and wrought iron regained its traditional position as the main end-product of the industry, establishing itself as a major export good for the first time. Output in south Wales increased to perhaps 40 per cent of the total, largely at the expense of those districts which had traditionally concentrated more on foundry work, Shropshire and, to a lesser extent, Derbyshire (15.3). In the 1830s and 1840s the industry underwent further restructuring, partly because of technological innovation, notably the 'hot blast' iron furnace in which air was heated above atmospheric temperature, and partly because of the advent of a major new customer, the locomotive railway. The introduction of hot blast had most impact in Scotland, where the industry expanded dramatically until by 1850 over a quarter of all pig was produced there, cutting the market shares of the Black Country and south Wales (where in both cases actual output rose) to about the same proportions. Three regions thus accounted for about 75 per cent of total output, while the remainder was fairly evenly distributed between Shropshire, Derbyshire and the West Riding, plus

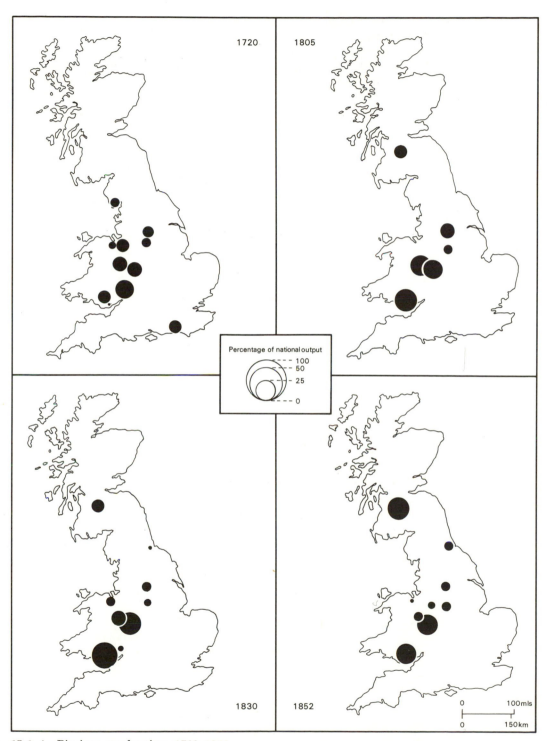

15.1–4   Pig iron production, 1720–1852

one new arrival, the Durham, Northumberland and Cleveland region (15.4). Domestic railway construction greatly increased the demand for wrought iron for locomotives and puddled and rolled iron for rails, of which south Wales became the most important producer. After 1850, overseas demand for railway iron was more important than that from British companies and helped to raise the proportion of exports to perhaps as much as 50 per cent of total output by the 1870s.

There was substantial relocation of the industry following the adoption of first the Bessemer process of steel-making, patented in 1856, and then the 'basic' process of 1879 (15.5–6). By 1870 the dramatic rise of the northeastern district, accounting for over a quarter of total output, was already apparent. The three older districts – south Wales, Scotland and the Black Country – all declined in the face of this competition, while at the same time other new iron-making centres were established elsewhere on the Jurassic ore-fields in Northamptonshire and Lincolnshire. These were some distance from supplies of fuel, so that for the first time coal was moved to where ore was to be found. By 1911 the concentration of steel production in the northeast, now responsible for more than a third of total output, was even more marked; not since the dominance of south Wales in the early nineteenth century had there been such dependence on one area. Then, south Wales had been one of three main districts; now only Scotland, with about 15 per cent of output in 1911, approached Cleveland's share. The other 50 per cent of output was divided between areas such as the Black Country and south Wales, where production survived on a much reduced scale; in Nottinghamshire and Derbyshire, where the industry enjoyed a modest revival; and in the newer districts of the northwest (Cumberland, Lancashire and northeast Wales) and the Midlands.

The restructuring of the industry in the later nineteenth century can also be seen in the rather limited statistics of iron ore output collected from 1855 onwards. In the 1850s (15.7) ore used in Britain was mostly mined in the coalfields close to the main centres of production. Thus south Wales, Scotland and the Black Country accounted for about three-quarters of pig output and almost the same proportion of ore, with the rest coming largely from Cleveland and Furness. In 1911 (15.8) the position was quite different. Imports of ore (in the 1850s amounting to some 20,000 tons a year compared with domestic output of 10 millions) were now responsible for 6.5 million tons out of a total consumption of 22 millions. Iron ore mining had virtually come to an end in south Wales and had greatly contracted in Scotland and the Black Country. Cleveland produced about the same proportion of the country's ore as it did pig (between 35 and 40 per cent) (Riden, 1980), while Cumberland and Lancashire each accounted for about a tenth. The most marked change was the rise of iron ore mining in the Jurassic belt of the Midlands between Oxfordshire and Lincolnshire, which produced nearly as much as Cleveland but largely for consumption outside the region in which it was mined.

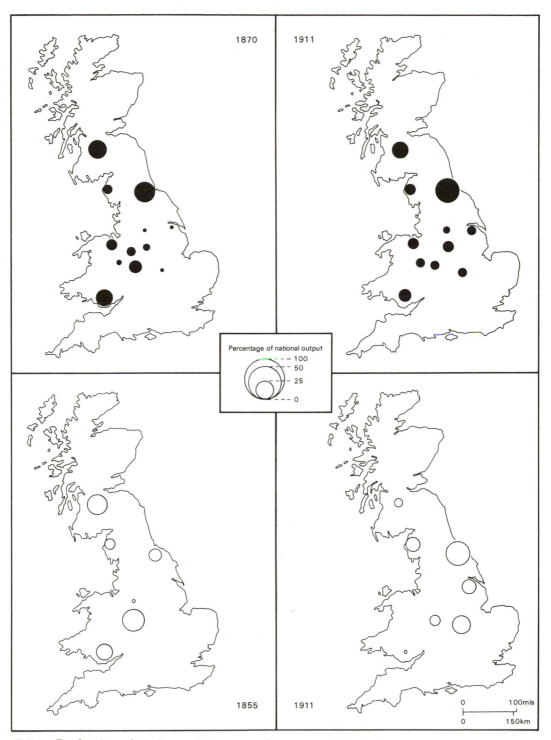

15.5–8   Production of pig iron, 1870–1911, and iron ore, 1855–1911

16 Shipbuilding

Anthony Slaven

Production and employment data outline the development of shipbuilding during the nineteenth century. Figures for tonnage constructed were compiled from the Navigation Accounts published in House of Commons Papers. Difficulties arise from changes in the calculation of tonnage, but here all tonnage is net tonnage of merchant vessels. This is a measure of cubic capacity of the vessel after deducting the space occupied by the crew, stores, fuel and machinery. A convenient series of aggregate output data may be found in Mitchell and Dean (1962). To achieve consistency employment data were compiled from the census. As used here they represent males in shipbuilding. These data are a less reliable guide to the changes in scale and distribution of the industry than the tonnage data owing to changes in classification and under-recording (Lee, 1979).

Shipbuilding experienced great changes in scale, location, technology and product in the nineteenth century (Slaven, 1980; Pollard and Robertson, 1979). The industry was not at the centre of the industrial revolution but came to play a crucial role in particular regions which developed an interdependent complex of coal, iron, steel and engineering trades focusing around shipbuilding. The output and employment maps are designed to illustrate these changes.

The three maps of shipbuilding tonnage, 1820 to 1911 (16.5–7), show the industry in its traditional form of wood and sail (1820), in transition to iron and steam (1871), and at the full development of modern steel shipbuilding (1911).

The 1820 pattern represents the scale and distribution of the traditional industry prior to the rise of steamships (16.5). Note the dominance of the Thames as the main area of construction. There was a widespread distribution over all rivers, with a heavy concentration between the Thames and the Bristol Channel. In the north the Mersey was important, reflecting its major port status, while the second major node of production of the northeast coast was linked to general European trade, but more specifically to the London coal trade. Scotland was a small producer and the east coast was more significant than the Clyde.

By 1871, output had increased sixfold from 65,018 tons to 382,585 tons. There was a dramatic switch of location to the northeast and particularly to the Clyde in Scotland (16.7). The districts in most rapid decline retained a large share of wood and sail construction, while the growth areas concentrated on iron and steam. Cheap malleable iron plates, the screw propeller and the compound marine steam engine lie behind these locational and constructional changes.

In 1911, output was now at 999,091 net tons, a scale increase of over two and a half times. The relocation of the industry had now been completed, only 6 per cent of construction lying outside the two main areas of the Clyde and northeast coast (16.6). The residual industry in the south coast areas and Thames remained tied to sail construction.

The maps of employment from 1831 to 1911 broadly confirm the changes in location and scale (16.1–4). In 1831, the Thames and south coast areas dominated. The northeast does not appear so important here as in the production data. The 1851 and 1871 maps indicate that the Thames remained more important in terms of employment for longer than the tonnage data suggest, and confirm the emergence of the Clyde from 1851. The 1911 pattern confirms the shift to Scotland and the northeast.

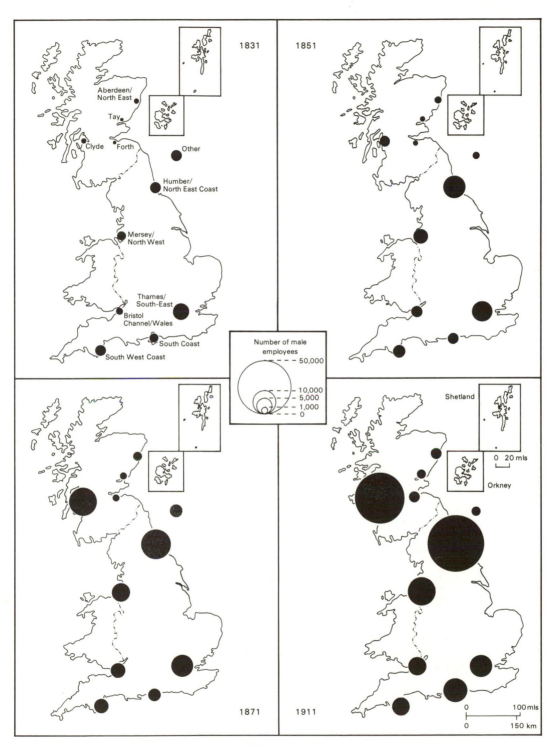

16.1–4 Shipbuilding employment, 1831–1911

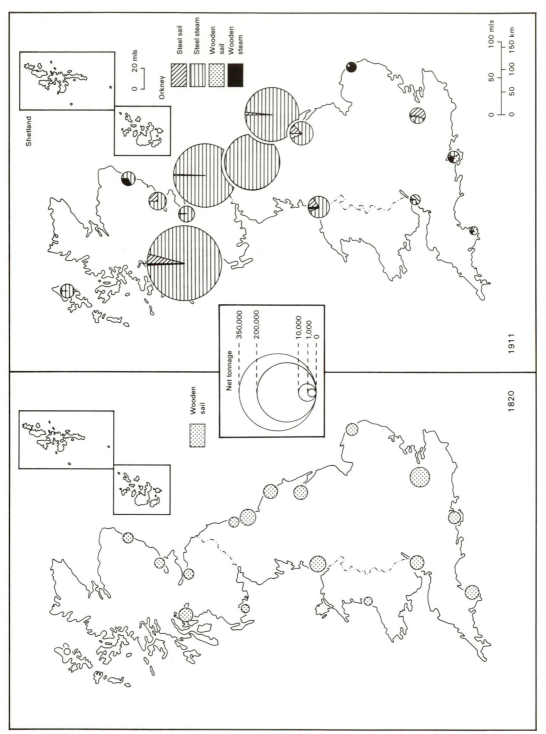

16.5-6   Tonnages of ships built, 1820 and 1911

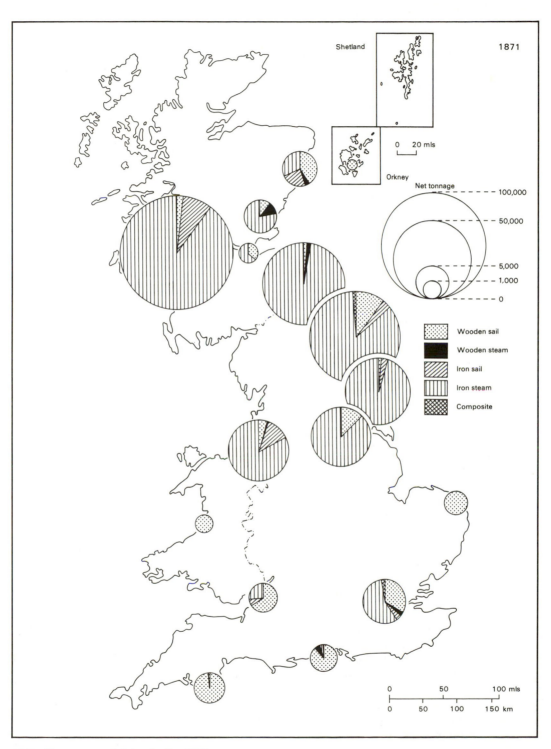

16.7   Tonnages of ships built, 1871

17 Engineering

John R. Hume and Miles Oglethorpe

Engineering was a vital force in industrialization but, before 1851, the first census which gives detailed occupational information, engineering seems to have been important neither as an employer nor as a consumer of capital. The 1851–1911 period saw development in depth.

Pre-1851 impressions gleaned from descriptive accounts allow fair guesses about major centres of engineering. The Black Country and south Yorkshire were well established as metal-working areas, but engineering developed in the manufacturing towns. Manchester, Leeds, and to a more limited extent, London, Birmingham and Tyneside were clearly important before 1820, with Glasgow, Greenock and Dundee in Scotland, and smaller towns in Lancashire and Yorkshire developing. Critical products of 'first generation' engineers were prime-movers (steam engines and waterwheels), transmission machinery (shafts and gearing), textile machinery, and chemical and mining plant. Much textile machinery of all types could, however, be made in small workshops, and chemical and mining plant, requiring little machining, could be made in iron and brass foundries, which also served architectural and other non-engineering customers. Companies which gained notable reputations in this early period included the Coalbrookdale Company (Shropshire), Boulton & Watt (Birmingham), the Carron Company (Falkirk), and Fenton, Murray & Wood (Leeds). Two of these concerns – Coalbrookdale and Carron – were predominantly iron-makers.

Textiles, mining and chemical machinery continued to account for a substantial part of total demand for engineering products throughout the early period, but the massive expansion of engineering activity that characterized the post-1820 period was based on new types of product, which in turn affected the location of producers. Railway locomotives and materials, marine equipment (including iron hulls, which are here classed as engineering products, by analogy with boilers and structural ironwork), and primary processing machinery for overseas, such as sugar, vegetable oil and rice machinery, were characteristic products post 1820. Most of these products were especially suitable for location and manufacture in ports, and London, Newcastle, Liverpool, Glasgow, Greenock, Leith and Dundee all became more important in this period. Although much railway engineering was concentrated in these towns, nodes of railway operation, emerging from the late 1830s, created opportunities for rational location of central workshops in areas once remote from large-scale economic activity, like Swindon and Crewe.

Underlying the growth trend was the increasing use of machines to make machines. The use of hand tools to shape wooden patterns for iron castings, and to file and to cut, became less dominant. Instead, 'steam arms' capable of tireless and accurate work took over more and more from skilled craftsmen. The prototypes of these new machine tools were devised in the first half of the nineteenth century, but their numbers increased enormously after 1850. They were essential for mass production, and were employed in their hundreds of thousands in factories set up from the 1860s to make sewing machines, bicycles, small arms and motor vehicles. The use of machine tools made components so much cheaper that they expanded the market for machinery of all kinds. The effect on location was threefold. First, certain areas like Manchester, the West Riding of Yorkshire, the Glasgow area and the West Midlands concentrated on tool-making (17.1 and 3). Secondly, new centres of machine-making emerged, using the new mass-production techniques, like Birmingham (small arms), Coventry (bicycles), Clydebank

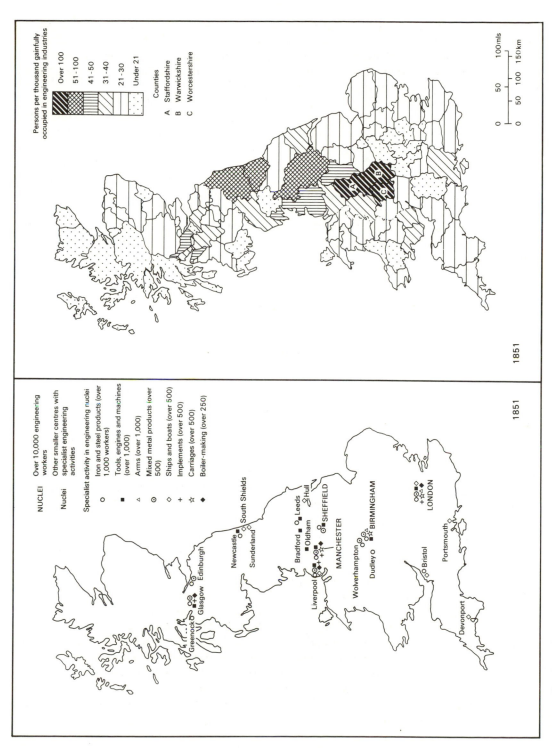

17.1–2   Engineering employment, 1851

(sewing machines) and Lincoln (agricultural machinery). Thirdly, the new machines allowed more complex equipment to be produced, such as mechanical harvesting and tilling devices, making machines a part of life for an ever-increasing proportion of the population. The late nineteenth-century appearance of the internal combustion engine heralded a major diffusion of engineering in Britain.

Advances in electrical engineering benefited a wide range of cities, especially the railway towns and some London suburbs. New methods of killing people came to the fore, ranging from the torpedo and the 14-inch naval gun to the machine-gun and the precision rifle. The smaller weapons could be mass-produced. The larger ones required massive new shops built in existing heavy engineering centres. Metal-working too became much more mechanized, with machines such as rolling mills, forging presses and mechanical manipulators of extraordinary size. Their output also found application in structural engineering, the making of bridges, buildings and piers from iron and steel, itself a new activity. Britain therefore had become a uniquely diversified and yet specialized engineering nation by the beginning of the First World War. There was hardly a branch of engineering production not represented in the country, and some were leaders in either European or world terms. This was reflected in an exceedingly complex structure of regional specialization, contrasting with firms making broad product ranges; of companies relying almost entirely upon exporting next to those catering for the home market; of hand techniques in the same concern as automatic or very heavy machinery.

The increasingly broad range of engineering activity is apparent from the maps which, by using data from the 1851 amd 1911 censuses, represent stages in the middle and towards the end of the period of the industrial revolution. Extracting the data for the diagrams was fraught with difficulties, most of which were attributable to the tendency for engineering to merge into many types of activity and, consequently, to the impossible task of defining engineering activity. It is for this reason that industries as diverse as shipbuilding and electrical goods manufacture have been absorbed into the data, resulting in the creation of a crude index of engineering activity rather than an absolute guide to the numbers of persons engaged in engineering throughout Britain.

Two types of map for each of the censuses have been constructed. The first (17.1 and 3) show the distribution of persons involved in engineering-related activity in absolute terms. Although these maps are useful, they tend to exaggerate the importance of major population centres such as London. For this reason, the remaining maps (17.2 and 4) are valuable because they represent the relative importance of engineering activity throughout the country by using the numbers of persons employed in engineering-related activity per thousand occupied persons.

The two sets of maps have produced a predictable pattern of engineering activity. By 1911 the engineering centres in the West Midlands, Tyne and Wear, Clydeside and the West Riding of Yorkshire had emerged. Although it is not possible to make direct comparisons between the 1851 and 1911 census data because of significant differences in occupation classification, marked changes in the pattern of engineering activity between 1851 and 1911 are clear. For example, the late and massive development of Clydeside is most dramatic, as is the extraordinary lack of engineering activity in Wales, where environmental conditions were not dissimilar to those of central Scotland.

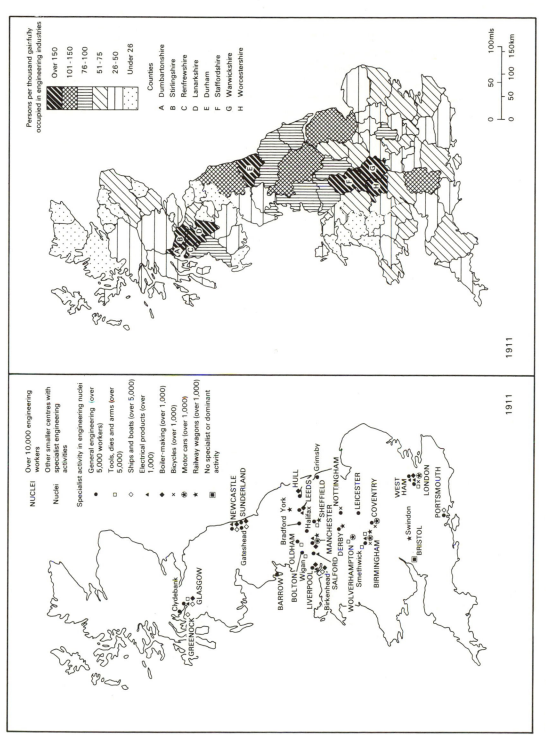

Persons per thousand gainfully occupied in engineering industries

Over 150
101-150
76-100
51-75
26-50
Under 26

Counties

A   Dumbartonshire
B   Stirlingshire
C   Renfrewshire
D   Lanarkshire
E   Durham
F   Staffordshire
G   Warwickshire
H   Worcestershire

1911

NUCLEI   Over 10,000 engineering workers

Nuclei   Other smaller centres with specialist engineering activities

Specialist activity in engineering nuclei

● General engineering (over 5,000 workers)
□ Tools, dies and arms (over 5,000)
◇ Ships and boats (over 5,000)
▲ Electrical products (over 1,000)
◆ Boiler-making (over 1,000)
✕ Bicycles (over 1,000)
✳ Motor cars (over 1,000)
★ Railway wagons (over 1,000)
■ No specialist or dominant activity

NEWCASTLE
SUNDERLAND
Gateshead
Clydebank
GREENOCK
GLASGOW
BARROW
BOLTON OLDHAM
Wigan
LIVERPOOL
Birkenhead
SALFORD MANCHESTER
WOLVERHAMPTON
Smethwick
BIRMINGHAM
Bradford York
Halifax LEEDS
SHEFFIELD
DERBY
NOTTINGHAM
LEICESTER
COVENTRY
Swindon
BRISTOL
HULL
Grimsby
WEST HAM
LONDON
PORTSMOUTH

1911

17.3–4   Engineering employment, 1911

18 Services
Clive Lee

Throughout the Victorian period, employment in the service sector increased rapidly, so that its share of national employment increased from a little over one-quarter in 1851 to almost 40 per cent by 1911 (Lee, 1979). Services include a wide variety of activities whose growth was stimulated by different social and economic forces. Some were primarily a response to increasing population, such as transport and distribution. When population growth is accompanied by increasing concentration in major cities as in the nineteenth century, engendering the scale effects of very large urban centres, there is a necessary growth in these services. One effect of large markets and large-scale production was an increasing separation between production and distribution. This was apparent in Victorian Britain in the food industries, in which processing became increasingly a separate activity rather than, as hitherto, the preserve of the grocer. Such activities were found principally in large cities and urban concentrations.

Other stimuli generated increases in other services. An increasingly complex and sophisticated economic system created a growing need for a wide range of commercial and financial services, from banking, insurance and stockbroking at one extreme to travelling salesmen at the other. Similarly, a substantial growth in professional occupations was a feature of the nineteenth century, including lawyers, clergy, the expanding teaching profession, and medicine (18.1–2). In part, of course, the expansion of such activities indicated an increasingly wealthy society and the service sector most obviously indicative of this was personal services (18.5–6). This included restaurant and hotel workers, and theatrical entertainers of all kinds. But this group was comprised overwhelmingly of domestic servants, especially female (18.3–4) in the nineteenth century. Not only was this group of workers the largest single employment category in the economy, with almost three million workers in 1911, it was more than twice the size of any other occupational group except textiles. The final service category includes government and defence, both of which grew substantially in response to the bureaucratic needs and military preferences of the nineteenth-century state. Employment in all the services grew considerably faster than employment as a whole. Miscellaneous services increased by a factor of 2, distribution and professions by over 2.5 times, transport and government by over 3.5 times, and the small banking and commercial sector of 1851 increased over thirtyfold by 1911.

The growth of the service sector was far from evenly spread throughout the country. Service employment per 10,000 by region showed very large variations. Certainly the creation and concentration of service employment was far stronger in London and the southeast than elsewhere. Within the context of a general increase in professions in the Victorian period, there was a much greater growth in the south, so that by 1911 only Somerset and the Lothians were in the same category as those counties in the immediate locality of London, which showed the highest level of concentration in professional employment. In the large personal service sector, Victorian growth also showed a clear relative advantage in the southern regions. The concentration of the south on services and the contrasting concentration of many other areas on manufacturing, reflect both different comparative advantages in the nineteenth century and also differences in income and wealth, services being concentrated in the most prosperous parts of the country (Rubinstein, 1981).

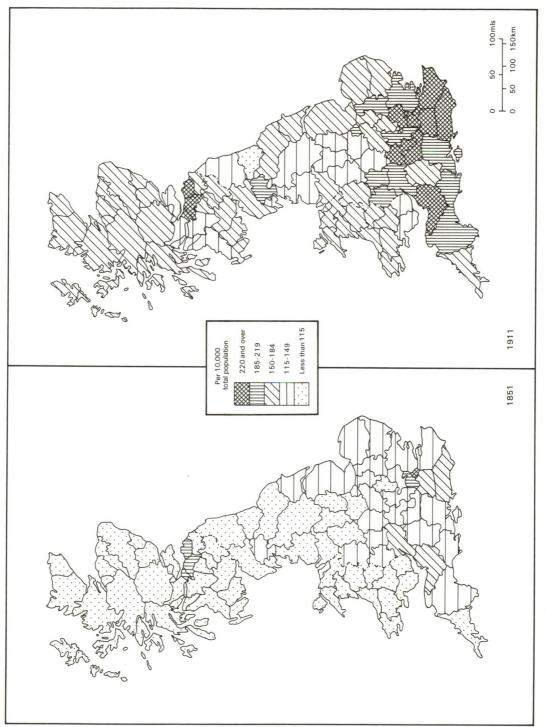

Per 10,000
total population

220 and over

185-219

150-184

115-149

Less than 115

1851

1911

18.1-2   Professional employment, 1851 and 1911

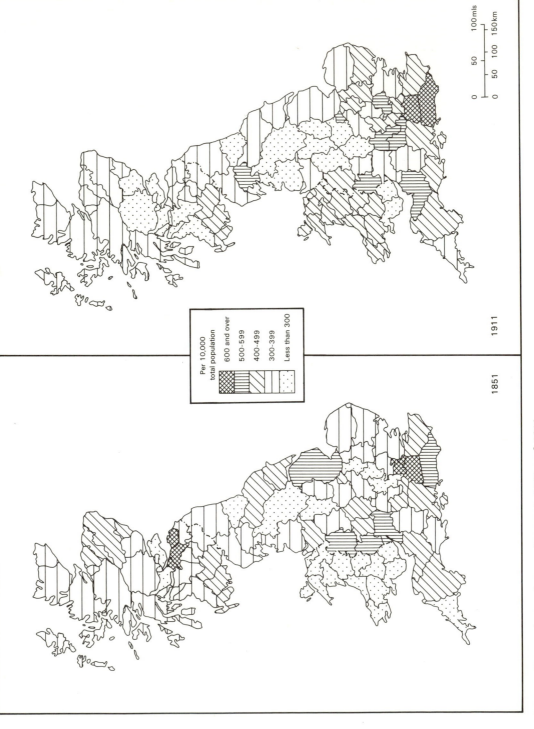

100 mls

100 km

1911

1851

Per 10,000
total population

600 and over

500-599

400-499

300-399

Less than 300

18.3-4 Domestic service employment, 1851 and 1911

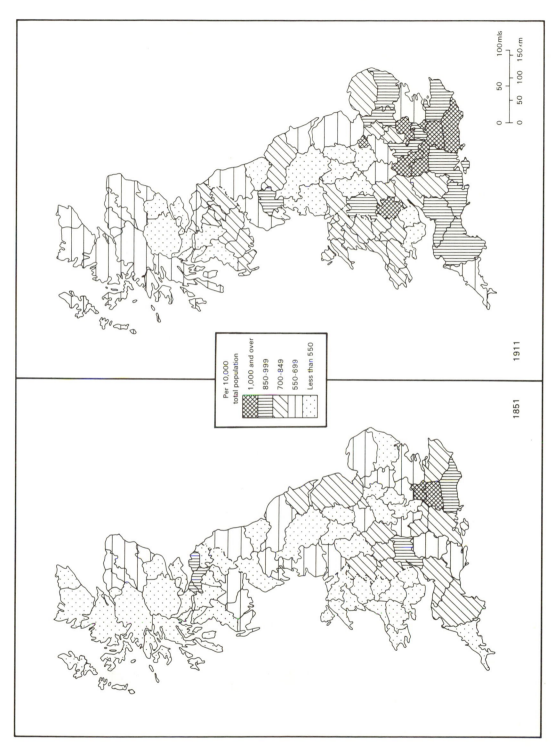

Per 10,000 total population

- 1,000 and over
- 850-999
- 700-849
- 550-699
- Less than 550

1911

1851

18.5-6   Personal service employment, 1851 and 1911

19 Banking and finance
P.L. Cottrell

Until the mid-eighteenth century, banking both north and south of Hadrian's Wall was a metropolitan business, centred in Edinburgh, Glasgow and London. In all three cities by the closing decades of the seventeenth century there were private bankers and proto-bankers, in London largely of goldsmith origins. Between the 1690s and the 1720s three public banks were established, the Bank of England in London in 1694, and in Edinburgh the Bank of Scotland in 1695 and the Royal Bank of Scotland in 1727. The Bank of England became a note issuer, the government's banker, and a provider of financial facilities to a select, largely metropolitan, private clientele. Its note circulation until the second quarter of the nineteenth century was confined mainly to the capital and the home counties. The Scottish public banks were also note issuers, took deposits at interest and, through the cash credit system, lent at short term. The private bankers were closely connected with the public banks but did not issue notes. Their chief liabilities were credits from the public banks together with some deposits.

In London at the end of the seventeenth century new private banks appeared, challenging the dominance of the older goldsmith houses. By 1800 there were nearly seventy in the City. The 'West End' group specialized in business with the gentry, aristocracy and lawyers, whilst the 'City' group dealt in commercial paper and the business of stockbrokers and country banks (Joslin, 1954; Cope, 1942).

The financial dimension of the industrial revolution was the emergence of provincial private banking, the numbers of such English and Welsh institutions being charted in 19.1. In 1750 Edmund Burke thought that there were about a dozen banks in England and Wales outside London. Thereafter there was a considerable growth in numbers, especially during periods of easy credit such as the early 1750s, the mid-1760s and the early 1770s, so that by 1784 there were probably some 120 country banks in England alone (Pressnell, 1956). Many were established by shopkeepers, corn merchants and general merchants who previously had informally undertaken banking functions as an extension of their business. Others who formed banks in the third quarter of the eighteenth century were turnpike officials, local receivers of taxes, and country attorneys 'who were able to put to profitable use money that passed through their hands'. On the basis of fragmentary directory evidence it would seem that the increase in the number of banks became even more rapid from the 1780s, especially after 1784 and assisted by easy credit conditions in the early 1790s, with the result that there was something of the order of 280 country banks in existence in England and Wales by 1793. The official series begins in 1809, following the Act of 1808 which required a licence for each bank office. Numbers fluctuated with commercial fortunes, but with the crisis of 1825–6 the number of country banks in England and Wales was sharply cut down.

Although the growth in the number of English and Welsh banks during the three decades after 1780 was substantial, it was but a catching-up of developments north of the border, as in the early 1770s Scottish banking was the most developed in Europe (Checkland, 1975). In terms of organizational and business practice Scotland was to remain in the van until the 1820s. During the 1760s, Glasgow and the southwest of Scotland was the initial centre for the expansion of provincial banking companies. This region remained the core when this growth reached a peak in 1810, when there were 25 provincial banking companies (19.4). The other major geographical concentration was in Tayside and Fife where the number of provincial banks rose from 2 in 1772 to 3 in 1792, 7 in 1810 and 8 in 1825. The only area of contraction in the Scottish industry was the

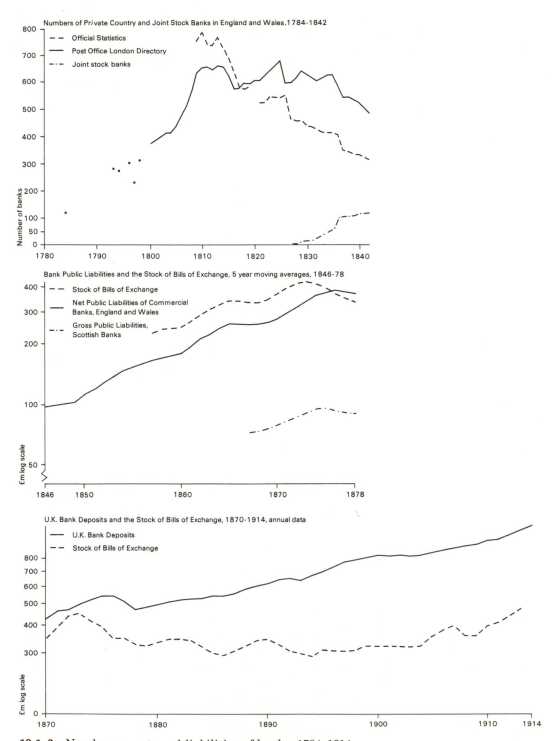

19.1–3 Numbers, assets and liabilities of banks, 1784–1914

decline of metropolitan private banking with the number of such firms in Edinburgh and Glasgow falling from nineteen in 1772 to ten, by 1810.

A distinguishing characteristic of British banking within Europe was unregulated private note issue and, given the interrelationship between note issue and the granting of cash credits, the issue of notes was of even greater importance to a Scots banker. In England during the last quarter of the eighteenth century and the opening decades of the nineteenth century, country bankers filled the gap in the circulation 'left by the failure of the mint and of the Bank of England to supply money for everyday needs'. The regional exception was Lancashire where small bills of exchange were used in place of notes (Ashton, 1945). There are indications that bills were important as a circulating medium in some other areas, such as the Black Country and the north Midlands. The heyday of the importance of English private notes for small payments was the first quarter of the nineteenth century. Licences to issue notes were first required in England and Wales in 1808 and in that year were taken out in every English county and all but one Welsh county, the exception being Radnor (19.5). Some note issuers in 1810 were industrialists rather than fully fledged bankers; some licences were obtained by bankers as a precaution, while many banks had very small note circulations. Consequently the county-by-county distribution of note-issuing licences in 1812 has to be viewed with a little caution. However certain features do clearly stand out; the importance of the industrial regions, but also the concentration in the south and west and in Yorkshire, marking the agricultural circulation.

After 1825 private banking and the private note issue in England and Wales became increasingly a feature of the agricultural counties. This was a result of the series of challenges which faced the English private banker after 1826 – the prohibition of notes under £5, those most needed to pay wages, the competition of the newly authorized joint-stock banks, and the growth of the Bank of England's own branch network in the provinces (Clapham, 1944). These changes in the institutional environment resulted in the number of note-issuing private bankers falling from 430 to 270 between 1833 and 1840. Following the 1826 Act the Bank of England under Horsley Palmer rapidly developed a provincial branch network to improve the quality of country circulation, with 2 branches being opened in 1826, 6 in 1827, 2 in 1829, and a further 2 in 1834. As Map 19.6 portrays, by 1837 the most important centres of the Bank's provincial note-issuing were the new industrial towns – Manchester, Liverpool, Birmingham, Newcastle and Leeds – an expansion partly arising from the Bank giving preferential account arrangements to provincial institutions which gave up their circulation and issued in their place Bank of England notes (Collins, 1972). The challenge from the new joint-stock banks came a decade later, with the most rapid growth in their numbers occurring in the mid-1830s, as shown in 19.1. The impact of these two developments is portrayed in Map 19.7 which reveals that by 1837 private bank notes were only of importance either within the Bank of England's 65-mile metropolitan radius, as with Buckinghamshire, Hertfordshire and Oxfordshire, or in predominantly agricultural counties such as Berkshire, Dorsetshire, Herefordshire, Lincolnshire and the East Riding of York-shire. This map also shows the continuing unimportance of notes in Cheshire and Lanca-shire and the inroads made by the Bank of England's post-1826 provincial circulation in such counties as Durham and Northumberland, Gloucestershire and Warwickshire from its branches at Newcastle, Bristol, Gloucester and Birmingham. In 1844 the Bank Charter Act halted the further formation of note-issuing banks, private or joint

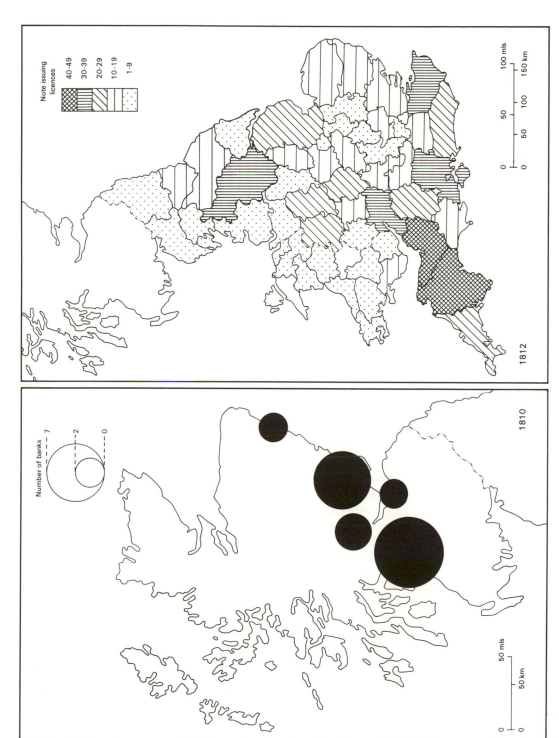

19.4–5 Provincial banks in Scotland, 1810; note-issuing licences in England and Wales, 1812

stock. Further it froze the existing level of the private circulation and established mechanisms for the eventual concentration of the note issue of England and Wales in the hands of the Bank of England (Collins, 1983). Actually, this legislation only accelerated an already existing tendency within English banking which from the late 1830s was shifting away from a reliance on note issuing. Initially the new joint-stock banks had gone into the business with 81.5 per cent of those formed before 1837 acquiring a note-issuing licence, but only 15 per cent of those established between 1837 and 1844 took this step.

Scots bankers protested against £5 being the smallest denomination for notes, which would have caused a severe curtailment of the cash credit system, at least for the provincial banking companies. This strength of feeling led to a parliamentary inquiry with the result that Scotland kept its small notes after 1829. Despite the vigour of the Caledonian opposition in the mid-1820s, note-issuing in Scotland was actually declining in terms of bankers' overall liabilities, as in England, with a fall from 24 per cent of Scottish banking liabilities in 1802 to 13 per cent by 1825. From the 1820s notes were only of importance to the Scottish provincial banks. The public banks and the Scottish joint-stock banks which grew in numbers during the 1830s came to concentrate upon deposit collection.

The small-note legislation of 1826 was part of Parliament's concern for financial and monetary stability. During the 1825–6 crisis eighty English private country banks failed, as opposed to two in Scotland, a contrast in experience which was exploited by Thomas Joplin in his campaign for the introduction of joint-stock banking in England. However, in his polemical pamphlets, Joplin did not compare like with like when referring to the Scottish banks, basing his arguments upon the few public and joint-stock concerns. Actually the Scottish provincial banks were either private partnerships or moderately sized co-partneries, similar to their southern counterparts, and only 19 of the 45 provincial banks formed between 1747 and 1836 in Scotland had more than thirteen shareholders. In general they were almost as failure-prone as English private banks, with a fifth meeting this fate before 1830 as compared with a third of English private banks. But what caught the public's and the legislator's eye was that only two Scottish banks failed during the financial turmoil of 1825–6. Consequently in 1826 an Act was passed which allowed the formation of note-issuing joint-stock banks in England and Wales outside a 65-mile radius of London. By 1837 there were 114 such institutions, falling back to 107 by 1844 when there were 286 private banks. In the mid-1840s the private banks had a bank:office ratio of 1:1.3 and the joint-stock 1:45, the latter ratio including the network of the National Provincial. Only Joplin's National Provincial attempted to implant the best of Scottish practice in terms of capitalization, organizational structure and management. In 1836 the National Provincial had thirty-nine widely scattered branches, each generally overseen by a local board and run by a manager and an accountant. The whole system of branches and agencies was watched over by two inspectors through annual visits. This was in sharp contrast to the behaviour of most of the other new English joint-stock provincial banks which generally were very similar to their private counterparts in terms of resources, management and branch networks. Largely untouched by Scottish innovative ideas, most English provincial joint-stock banks remained rooted in the local, sometimes regional, economies that they had been established to serve, until the last quarter of the nineteenth century. This structure is shown in Map 19.8 which depicts the distribution

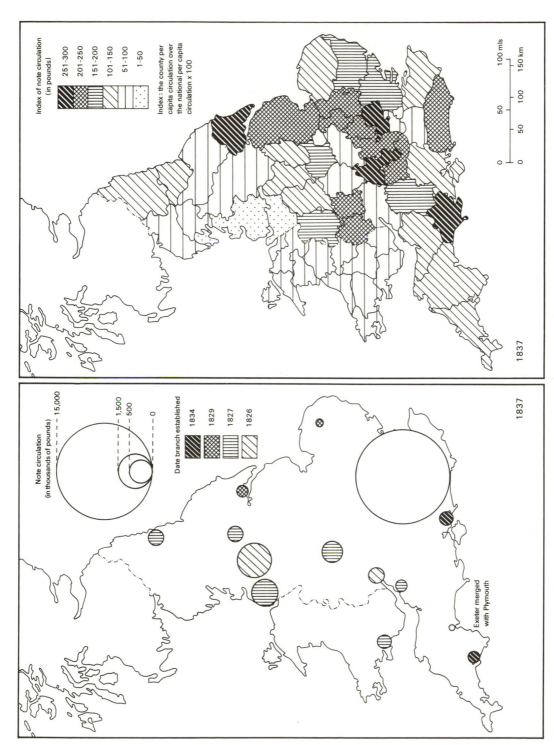

19.6-7 Note circulation: Bank of England and private country banks, 1837

of provincial joint-stock bank offices in England and Wales in 1842. The metropolitan radius clearly stands out with the only joint-stock bank offices in the southeast being in the furthest eastern tip of Kent. The other contrast is the national network of the branches of the National Provincial and the regional distribution of the offices of the North and South Wales Bank, compared with generally the very parochial spheres of business interest of the other banks.

Although English provincial joint-stock banking for half a century after 1826 did not change in nature and quality, Scottish ideas made great strides in London following the declaratory clause of the Bank Charter Act of 1833 which allowed the formation of deposit-taking joint-stock banks within the metropolitan radius. By 1836 there were three London joint-stock banks and by 1857 eight, all deliberately formed to draw upon the savings of the metropolitan middle class through the payment of interest upon term deposits.

When Joplin had mounted his campaign for the introduction of joint-stock banking in England in the mid-1820s, there were only three non-chartered joint-stock banks in Scotland, of which two were still in the process of formation. This development had started with the establishment of the Commercial Bank of Scotland in 1810 in Edinburgh. Its promoters, critics of the city's public banks, had aimed to establish a new institution on the basis of a very large group of socially diverse investors as opposed to the small wealthy cliques which backed the public banks. They were successful, with the Commercial having 673 shareholders within the first year of its existence. A decade and a half later the National Bank, the fusion of three rival schemes, was also established on a similar basis in Edinburgh, having 1300 shareholders. In the same year, 1825, the Aberdeen Town and Country Bank was formed as a regional joint-stock bank for the northeast. However, until 1830, despite the growth of the provincial banking companies after 1760, Edinburgh with its three public banks and two joint-stock banks, remained the financial centre of the Scottish economy. Banking in Glasgow, the industrial centre, remained relatively backward in Scottish terms until the 1830s when four joint-stock banks were founded. This change in Glasgow, a catching-up of financial growth and organization with industrial developments, was mirrored elsewhere, joint-stock banks being established in the 1830s in Aberdeen, Perth and Dundee. These new institutions usurped and replaced the private banks and the provincial banking companies.

In 19.2 the growth of English bank liabilities and the stock of bills is charted for the mid-century period. The trace is based upon estimates of the amount of notes and deposits drawn from available bank balance sheets and adjusted to be representative for the whole banking sector. It aims to portray the medium trend and it is reinforced here by being plotted in a semi-logarithmic form. As the diagram shows, bank liabilities, largely deposits, increased at about 4.1 per cent per annum from the late 1840s to the late 1870s, and accordingly were growing relative to both population and Gross National Product. This rate of growth largely reflects the experience of the joint-stock banks whose net public liabilities were increasing at about 5.9 per cent, rather than that of the private banks for which notes and deposits were increasing at only 2.2 per cent. Although the estimate is of the medium-term trend, it is marked by fluctuations, with two sharp contractions following the banking and liquidity crises of 1847 and 1866 and a falling away in the mid-1870s with the growing deflationary climate of that decade. However, the early 1850s and mid-1860s were periods of sharp expansion, with the

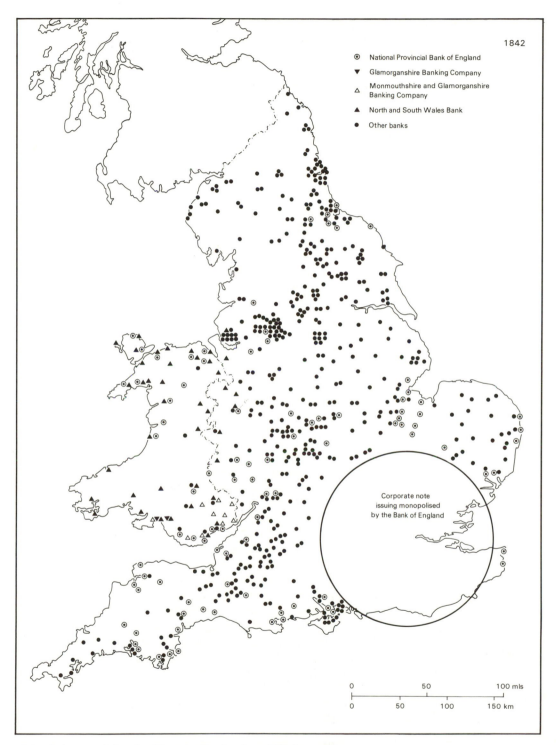

Legend:
- ⊙ National Provincial Bank of England
- ▼ Glamorganshire Banking Company
- △ Monmouthshire and Glamorganshire Banking Company
- ▲ North and South Wales Bank
- ● Other banks

1842

Corporate note issuing monopolised by the Bank of England

19.8 Joint-stock bank offices in England and Wales, 1842 monetary impulses of the 'new' gold and the reform of company law which led to further waves of bank formation in the 1860s and early 1870s. The estimate of the gross public liabilities of the Scottish banks for the period from the mid-1860s to the late 1870s shows a similar pattern of fluctuations to that of the English series, with a pause in growth following the 1866 crisis, a cyclical boom in the early 1870s and a decline from the mid-1870s.

Data on the volume of bills of exchange have been converted into a stock estimate, assuming a usance of three months, for comparability (Nishimura, 1971; Capie and Webber, 1985). This series displays a strong parallelism to those for bank liabilities, with cyclical expansions during the booms of the 1860s and the early 1870s. The other important feature is the cross-over point in the mid-1870s. The estimate of the stock of bills is an indication of the demand for credit in the economy, and with bills exceeding bank deposits until the mid-1870s, the financial sector was dangerously poised with many banks in industrial areas 'overlent'. The consequence was that the banking system was vulnerable to any shock. However, with transport improvements, both domestic and international, inventories and the resultant need for bill finance declined from the mid-1870s. With deposits growing relative to advances and discounts banks now had greater liquidity and were thereby able to accommodate their customers by overdraft facilities rather than through discounting bills.

These trends can be followed through for the late nineteenth century in 19.3 which displays the stock of bills falling away from the mid-1870s until the mid-1890s but with cyclical fluctuations. The recovery of the stock of bills during the twenty years before the First World War was largely connected with international factors, in particular the role of finance bills in the operations of the gold standard. Although the estimate of bank deposits plotted is in the form of annual data, overall its growth path is smoother than the five-year moving average series for the mid-nineteenth century. The only cyclical fluctuations which are clearly marked are the deflation of the late 1870s, compounded by the City of Glasgow Bank crash, and a short pause during the trough of the early 1890s.

The fragmented organizational structure of English banking until the late nineteenth century, arising from the typicality of local banks with few branches, resulted in integration occurring only as a product of the correspondent system. Map 19.9 indicates the geographical distribution of the country banks for which the London private bank of Glyns acted as agents in 1850. In this arrangement country banks were able to draw bills on Glyns and have their notes payable at Glyns, maintaining a balance with Glyns for these purposes. Glyns, a 'City' London private bank, had 4 country agency accounts in 1790, 13 in 1800, 38 in 1810 and 60 in 1849, by which time half of these were for provincial joint-stock banks. Initially Glyns had acquired these accounts through the personal and marriage connections of its partners. Through an account with a London correspondent, banks in credit surplus areas of the economy, largely the agricultural districts of the south and east after the Napoleonic wars, were able to send 'excess' deposits to London for investment in bills of exchange. Alternatively 'overlent' banks in the industrial areas sent bills for rediscount in order to maintain liquidity. From the 1800s the London money market became institutionalized with first the appearance of bill brokers, while from the 1830s there emerged at the centre of the market discount houses which in turn developed direct connections with the country banks in order to obtain both deposits and bills.

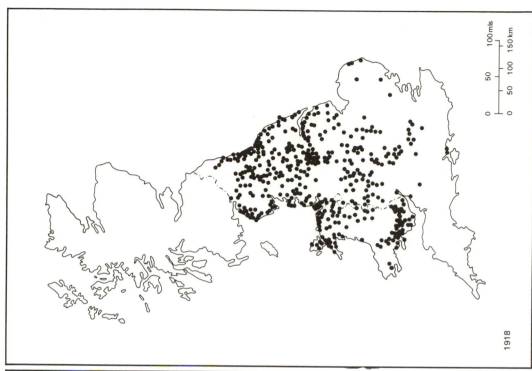

1918

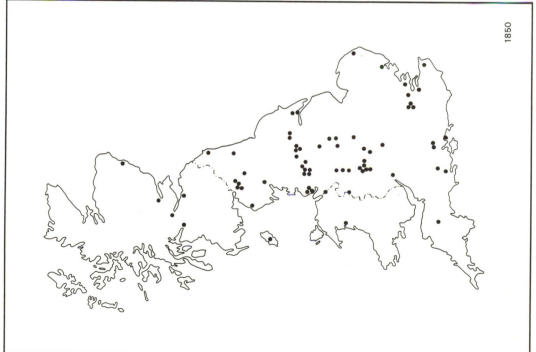

1850

19.9–10 Correspondent banks of Glyn Mills, 1850; constituent banks of the Midland Bank in England and Wales, 1918

The decline in the use of domestic trade bills from the 1870s and, more importantly, the amalgamation movement amongst banks led ultimately to English joint-stock banks having 'balanced branch networks' by the 1900s. This allowed banks to offset regional deficits and surpluses of liquidity internally. Bank mergers had occurred throughout the nineteenth century, although with something of a lull during the 1840s and 1850s. However, the rate of merger activity accelerated from the late 1880s and now involved the absorption of one joint-stock bank by another, instead of the acquisition of private banks by joint-stock banks which had been the previous pattern. It would seem that competition through size became the dominant characteristic of English banking in the last decades of the nineteenth century, with the lead being taken by first Lloyds and then the London and Midland. Between the 1850s and the 1890s the Midland had proceeded cautiously with respect to mergers, acquiring only a few banks in its immediate region around Birmingham – in Coventry, Leamington and Derby. However as Map 19.10 shows, from 1890 the Midland took vigorous action to become a national bank through taking over other joint-stock banks, first in Leeds, then in London, followed by acquisitions in the northwest, the West Riding, Leicestershire, Sheffield, Nottingham, and north and mid Wales. However, these amalgamations only increased the Midland's presence in industrial areas and consequently did not improve its overall liquidity position. The Midland only developed a 'balanced' branch network through these acquisitions going hand-in-hand with a policy of continuous branch expansion in southern England in the 1890s and in East Anglia after 1903. Mergers and the opening of new branches gave the Midland not only a national presence but also a high rate of growth of deposits: 4.6 per cent per annum between 1890 and 1914.

Banks from the mid-nineteenth century were largely, although not entirely, concerned with the provision of credit through the mobilization of deposits. The development of stock exchanges was the corresponding process of institutionalization of the capital market (Morgan and Thomas, 1962; Killick and Thomas, 1970). There was a securities market in London from the end of the seventeenth century, although the Stock Exchange was not formally established until 1802. Provincial developments came with the creation of joint-stock banks in the 1830s and the railway manias of the 1830s and the 1840s. In 1836 formal stock markets were established in Liverpool and Manchester, while during the mid-1840s nearly every large city and industrial town in the British Isles had a stock market of some form. In all, fourteen formal exchanges were established during the railway promotion boom of the 1840s, including those of Edinburgh, Glasgow and Aberdeen. However, with the fall of railway share prices from the autumn of 1845, most of these institutions collapsed through the loss of turnover, income and members (19.11). The London market continued to grow, particularly from the mid-1850s, while from the 1870s provincial markets re-emerged with the gradual adoption of limited liability by industry and commerce (19.12). The two most important periods of provincial recovery were the domestic investment booms of the first half of the 1870s and the mid-1890s as the re-established markets came to concentrate upon local industrial securities, with brokers having particular expertise and the ownership of the shares being often concentrated geographically. Ultimately domestic industrial and commercial shares and debentures were dealt in on the London market, but its main business lay in dealings in British government securities, railway stocks and bonds and overseas issues.

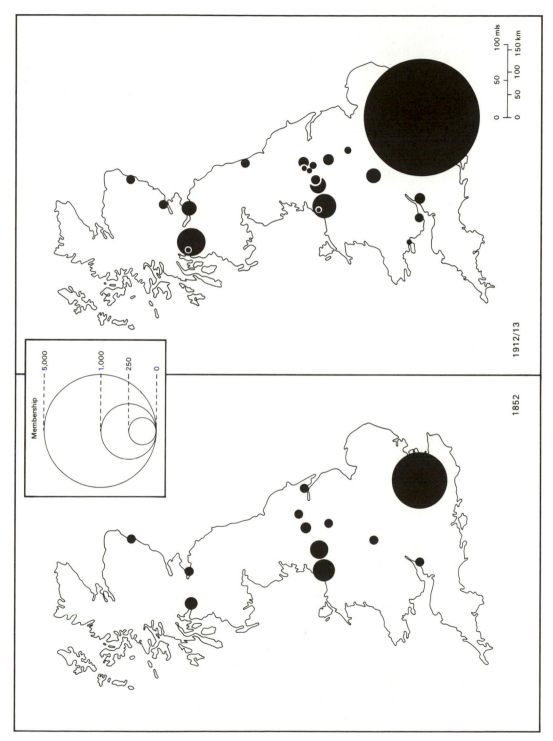

19.11–12 Stock Exchange membership, 1852 and 1912/13

Wealth and the wealthy

W.D. Rubinstein

This section examines the distribution of wealth and income among the wealthy and middle classes between the early nineteenth century and 1914. Maps 20.1–3 detail the main business venues of Britain's richest non-landed (business and professional) wealth-holders dying between 1809 and 1914. The names and wealth at death of these men are taken from the probate records of wealth left at death. 'Wealth-holders' are British males who left £500,000 or more between 1809 and 1914. By 'business venue' is meant the place in which the wealth-holder carried out his business activities, not the locale in which he may have lived or retired. The Rothschild merchant banking dynasty is clearly located in the City of London, and the Wills tobacco family in Bristol. By 'non-landed' is meant all males leaving £500,000 or more apart from bona-fide landowners – that is, landowners whose families were not *primarily* in receipt of a business or professional income for three generations or more. Since the genealogy and family histories of nearly all substantial landowners can be readily traced, there is very little ambiguity in this definition. Non-landed wealth-holders also include all scions of business families less than four generations from active participation in trade or the professions, who were not themselves active in business – that is, rentiers and the 'idle rich', as well as professionals like Lord Eldon, the Lord Chancellor, who acquired fortunes of £500,000 or more.

The geographical units here include all of Britain's major conurbations and larger cities, and larger areas containing the residue. London has been divided into the City of London, Britain's financial centre, Other London (the old LCC area) and Outer London (the outer GLC region). The curious dates delimiting the first map are a product of the sources. Probate values of very large estates originate on a comprehensive basis in England in 1809, while on 9 January 1858 the Civil Principal Probate Registry (situated since 1874 at Somerset House) superseded the old ecclesiastical courts in the probating of wills and administrations, its jurisdiction covering all of England and Wales. The Scottish data begin in 1825.

The maps underline the importance of London, especially the City, in the making of large British fortunes. It was important because it was the centre of Britain's commercial and financial activity. Outside London it was commercial centres like Liverpool which dominated. Only one Manchester cotton manufacturer left over a million pounds between 1809 and 1895. As the figures exclude wealthy landowners who made London their social centre and provincial businessmen who bought houses there, they understate the capital's wealth. There were several small pockets of wealth: Bristol with Wills tobacco, and Belper in the Midlands with the 'beerage', like Bass and Allsopp. By 1914, areas like the west Midlands and Clydeside had increased in importance (Rubinstein, 1977 and 1981).

Maps 20.4–6 survey the geographical distribution of taxable (i.e. middle-class) income at three dates between 1806 and 1911–12. Income tax was levied in Britain between 1798 and 1815, when it was abolished, and again from 1842. Anyone with a yearly income of £50 (sometimes £60) in the period 1798–1815, and of £150 (later £160) after 1842, was liable to pay tax on his income (Hope-Jones, 1939; Stamp, 1916). The income tax figures thus take in virtually the whole of the middle classes as they were commonly understood in income terms, while excluding all but some skilled wage-earners.

The figures indicated in Maps 20.4–6 are far from straightforward. Between 1803 and 1911, income tax was not simply levied on individual taxpayers and it is not possible to

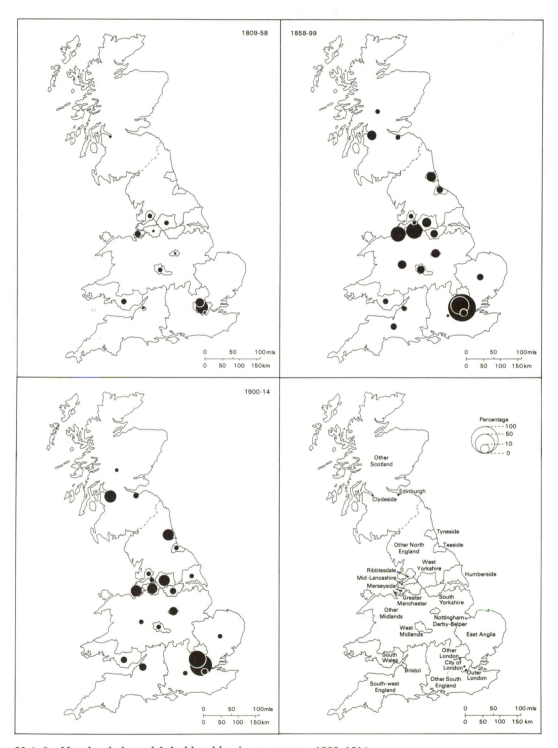

20.1–3   Non-landed wealth-holders' business venues, 1809–1914

say precisely how many individual taxpayers there were, let alone their geographical distribution. The maps here indicate the total county-by-county assessments for total income declared under three of five Schedules. Schedule D – the assessment on business and professional incomes – and Schedule E, that on the incomes of employees of corporations (including non-business corporations like Oxbridge colleges), are included in the totals, as are those assessments made under Schedule A, which separately detailed certain types of business activities, including mines, quarries, canals and railways. Tax on land rent, farmers' incomes and income from government securities are not included in these figures. The sources did not differentiate business units from individual taxpayers, and the figures in these maps also include, besides taxpayers, the tax on the profits of businesses. The data in Maps 20.5–6 are taken from the manuscript income tax returns (I.R. 16) currently held at the public Record Office, Kew, while the 1806 statistics (20.4) are taken from the county returns of income tax published in the Parliamentary Papers in 1815–16.

London is again dominant, paying 39 per cent of all British income tax in 1806 and 42 per cent in 1911–12. There was however a drop to 27 per cent in mid-century with the growth of the newer northern industries. There was a general decline in the proportion of tax coming from southwest England and East Anglia, and a relative decline in the share of the less urbanized counties. The wealth of the industrial north never caught up with that of the commercial, serviced-based middle classes of London.

Landed property was the most significant form of nineteenth-century wealth. It largely determined social status, accounted for half the members of all Cabinets before 1905, and was disproportionately concentrated in a small circle of landowners. Map 20.7 shows the distribution of the two highest categories of Bateman's summary of the 1872–4 Parliamentary survey. 'Great landowners' were landowners who owned 3000 acres or more and had a rent roll of over £3000. As most peers met these conditions, Bateman provides an accurate survey of inequality in landowning. London was not surveyed and remains a major gap. The degree of concentration was phenomenal. 1688 peers and great landowners owned 41 per cent of all land in England and Wales. Bedfordshire was owned by seventeen people. Ownership was least concentrated in the home counties, in East Anglia, and in Cumberland and Westmorland where there was a strong tradition of small-owner occupation. Concentration was highest in the northwest Midlands and in Northumberland.

Scottish nineteenth-century landownership was even more unequal, although the two sets of figures, derived from the same source, are not entirely comparable. Bateman's *Great Landowners* contains county tables of the distribution of landownership in England and Wales, taking as its criteria for the definition of 'great landowners' the ownership of 3000 or more acres if the rental reached £3000 or more. Since Scottish rentals were, on average, half of English values, only ownership of 3000 acres has been taken in defining a Scottish 'great landowner'. Thus there will be more Scottish 'great landowners' by these criteria than is the case in England, but the extreme concentration of landownership in Scotland was a fact none the less. The extremes were reached in Buteshire, where four aristocrats owned 96 per cent of the whole county, and in the more celebrated case of Sutherlandshire, where the Duke of Sutherland alone owned 91 per cent of the county.

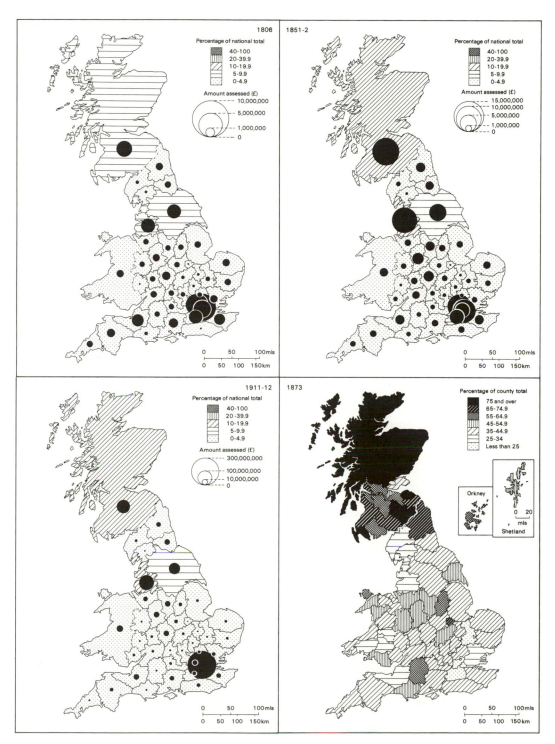

20.4–7 Income tax yields, 1806–1912; land owned by great landowners, 1873

21 **Poor law and pauperism**

Ian Levitt

In 1834 the English Poor Law was drastically and dramatically reformed. Parish-based administration was replaced by Boards of Guardians and 630 amalgamations of parishes. Although the new national controlling body was to change its name from the Poor Law Commissioners to the Poor Law Board and then the Local Government Board, the power it exercised heralded a new era of utilitarian management. The reform, a fusion between Malthusian fatalism and a *laissez-faire* ideology, was primarily aimed at control of the high levels of welfare spending that had become prominent during the Napoleonic wars. The Speenhamland system, although not the only form of subsidizing low-paid agricultural labour, was the most notorious, and by all accounts during certain periods of distress over 20 per cent of southern England had been on parish relief (Marshall, 1978).

Scotland, by contrast, had evolved a totally different system, largely based on the kirk, and more importantly, on voluntary giving. Although legal assessments (rates) could be applied, by the 1830s only the larger towns and the parishes in the southeast (where economic conditions resembled those in England) had introduced such a measure. The results of this system were to create a Poor Law in name only. Allowances, where given, were much smaller. Indeed in some places the title pauper was one of status; it allowed a person the licence to beg. The whole system was underlined by many parishes refusing to give assistance to the unemployed.

The rates of pauperism were varied in both Scotland (Cage, 1981) (1837) and England (1839) (21.1). Despite the anti-Speenhamland ideology of the Poor Law Commission and the growth of workhouse provision, a broad band of high pauperism from The Wash to Portland Bill existed. North Wales also had high rates. The lowest were across central England and in south Wales. It can also be seen that the rates in Scotland were almost uniformly below the English counties, even those that had legal assessments. The industrial areas of Scotland (21.4) had only half the rate of the English industrial counties, an underlying reason why the 1840s saw a clamour for the reform of its Poor Law. There was a need, it was argued, to provide the working classes with some form of systematized welfare. After their own Royal Commission the 1845 Poor Law Act was passed. It similarly brought in central control, through the Board of Supervision, but left the parish-based administration. The tone of the administration was, however, to be radically different; all those entitled were to receive adequate allowances (Fraser, 1976).

After the economic distress of the early 1840s the rate of pauperism in England declined from an overall average of 7 per cent to around 5 per cent (21.1). In Scotland, the Board of Supervision was quick to ensure the Poor Law's acceptance and the numbers assisted rose. Although economic distress in the next twenty years was to cause the rate of pauperism to vary, the 1869 map (21.2) confirms how far the two countries had moved towards relieving similar proportions. Allowing for the non-relief of the unemployed, the rates were virtually the same. The year 1869 was, however, a watershed. With many workers now reaping the rewards of sustained economic progress there seemed little prospect of unrest over the nature of welfare, and when new administrators took over in Scotland they began to argue for the implementation of a much tougher policy. To them it was important to distinguish between poverty, a regrettable by-product of economic growth, and pauperism, a social disease created by too much reliance on state support. If many amongst the working classes had improved themselves through hard work, thrift and independence, then it was up to the state to encourage others in the same path. The

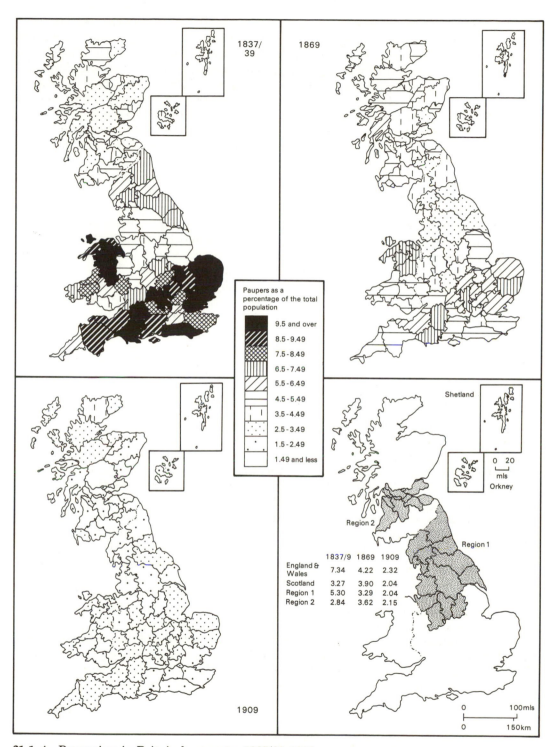

Paupers as a percentage of the total population

■	9.5 and over
	8.5 - 9.49
	7.5 - 8.49
	6.5 - 7.49
	5.5 - 6.49
	4.5 - 5.49
	3.5 - 4.49
	2.5 - 3.49
	1.5 - 2.49
□	1.49 and less

	1837/9	1869	1909
England & Wales	7.34	4.22	2.32
Scotland	3.27	3.90	2.04
Region 1	5.30	3.29	2.04
Region 2	2.84	3.62	2.15

21.1–4   Pauperism in Britain by county, 1837/39–1909

Poor Law was therefore seen as a natural instrument to encourage the pauper class. Through organizations like the Charity Organization Society this movement spread to England and by the mid-1870s thousands, particularly mothers with illegitimate children, had the workhouse test more ruthlessly applied. Indeed in both England and Scotland a new wave of workhouse building took place, leaving the poor in no visible doubt over the policy (Digby, 1981; Crowther, 1981). 1909 (the last year before elderly paupers received their non-contributory old-age pension and just two years before the introduction of national insurance) shows how successful that policy had been (21.3). The average rate of 2 per cent had in fact been achieved by the late 1880s. The Poor Law had become the antithesis of a welfare service.

Nevertheless the average yearly rate hides the flow of those seeking relief throughout the year and the English map of 1907 shows that the southern Unions assisted considerable numbers (21.5). Even allowing for the insane poor, some had in excess of 6 per cent on their rolls. The largest inflow occurred during the winter months when demand for agricultural labour declined. Speenhamland may have long been buried, but many Unions evidently felt statutory assistance remained a necessity. It should also be noted that the urban industrial Unions around Birmingham, Manchester, Liverpool and London also had substantial numbers seeking and receiving assistance, an indication that the growth of urban poverty had created a new form of distress. If the poor in these areas were actually willing to endure the rigorous application of the test, then the fear that the 1834 reformers had pleaded for, and the 1870 administrators had implemented, was of no avail.

The nineteenth-century Poor Law was a testament to the gradual evolution of a nationally implemented statutory welfare policy. Before 1830 local conditions, particularly in the south where there was a need for agricultural labour, produced varying practices. Whilst some parishes were prepared to assist substantial numbers, others, irrespective of poverty, operated as if the Poor Law was a legal fiction. Once administrative networks became established (1830s for England, 1840s for Scotland) then this divergence began to be curtailed. However, it was not until the 1870s, when the national administrations felt more secure, that any drive for full convergence occurred.

The rates for Scotland in 1837 have been calculated from returns supplied to the General Assembly of the Church of Scotland's Inquiry. They are based on actual numbers seeking relief throughout the year, but in some places the figures omit what was termed the *occasional poor*, a name often used to denote the unemployed. The county population has been adjusted for the small number of parishes that made no return. No one-day count took place until 1859. The rates for England in 1839 refer to a one-day count held on 25 March (Lady-day), the first to cover over 80 per cent of the population. It referred only to those Unions administering the 1834 Poor Law Amendment Act; returns were therefore most incomplete for Lancashire. The rates for 1869 and 1909 were calculated from the one-day count of 14 May in Scotland (15 May in 1909), and from an average of 1 January and 1 July in England. By then Scottish administrators collected data for 15 January and 15 September and the May figure was usually half way between the January high and the September low. In England the January total was usually the higher. The insane poor are excluded except in the 1907 map. The English counties are the Union counties.

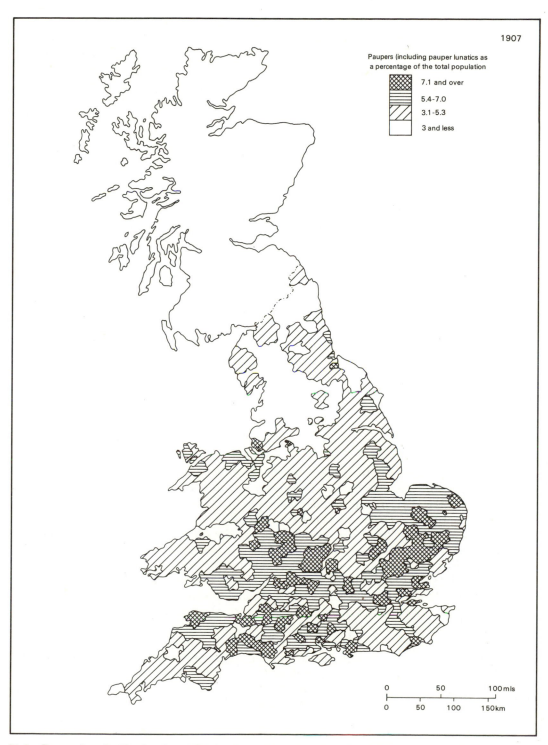

1907

Paupers (including pauper lunatics as
a percentage of the total population

7.1 and over

5.4-7.0

3.1-5.3

3 and less

0		50		100 mls
0	50	100	150 km	

21.5   Pauperism in England and Wales by registration district, 1907

22 **Urbanization**

R.J. Morris

Industrialization in Britain as in many other economies included not only a change in economic structure and a sustained increase in population and per capita income, but also changes in the spatial organization of that population which included both urban growth (an increase in the size of towns) and urbanization (an increase in the proportion of the population living in towns). Maps 22.1–3 show the location of the urban places involved in this change. These maps use data developed by C. M. Law and Brian Robson (1973) which accord well with the traditional definition of urban places as agglomerations of population of relative size, density, and heterogeneity. Three modifications have been made. Scottish data have been added from the printed census. The changing scale of urbanization has been recognized by changing the lower limit of the operational definition of urban from 5000 in 1801 to 10,000 in 1861 and 20,000 in 1911 (the modern UN definition of an urban place). The characteristic of towns as places for the organization of power emphasized by Max Weber has been recognized by re-dividing places like Manchester and Salford which had an independent political character. There is a natural and very real conflict between the geographical concept of contiguity and agglomeration, and the sociological and historical concept of independent political and social identities.

Maps 22.1–5 show clearly the increasing size and number of urban places. New entrants were places with a population below 2000 at the start of the time period. The growth rates were the simple mean annual increase as a percentage of the population at the start of the time period. The rank order size chart (22.4) plots size against rank order. If the relationship was one of proportionality (i.e. the second largest town was half the size of the largest, the third largest, a third the size and so on), then the results of the log scale graph would be a straight line. Although some older theorists asserted that such proportionality indicated a developed urban system (after all the USA was like that), such a standard should only be used for descriptive purposes. In this case the continued dominance of London is clear. The historical emphasis given to the provincial capitals was due to the newness of the 100,000 cities and their rapid growth. Any apparent decline in London was only in relation to previous massive dominance. In 1811 London was twelve times its nearest rival, in 1861 five and a half times, with a slight recovery to six times in 1911. Britain retained a metropolitan urban system throughout the nineteenth century, dominated by its primate city. There were other patterns. In the first half of the century growth was dominated by the resource-based textile and metal goods towns of Lancashire, Yorkshire and the west Midlands. St Helens, Burnley and Bradford grew at over 10 per cent a year. In addition there was a ring of towns around London like Luton and Enfield, and a group of leisure, residential and retirement towns, like Brighton, Cheltenham, Hastings and Torquay, mainly along the south coast, that attested to the strength of the service and commercial economy. The second half of the century witnessed growth in the capital goods and coal exporting areas of south Wales, northeast England and the west of Scotland. Clydebank, Jarrow, Middlesbrough, Cardiff and Mountain Ash represent this group. A large number of coal communities, like Ashington and Rhondda, were involved here. They have the size and density of urban communities but not always their heterogeneity. Perhaps they should be thought of as urban places, but without the means of organizing power possible in places more broadly based in history, culture and the economy. The ring around London thickened with places like Harrow, Woking and Watford joining the system. Such grouping

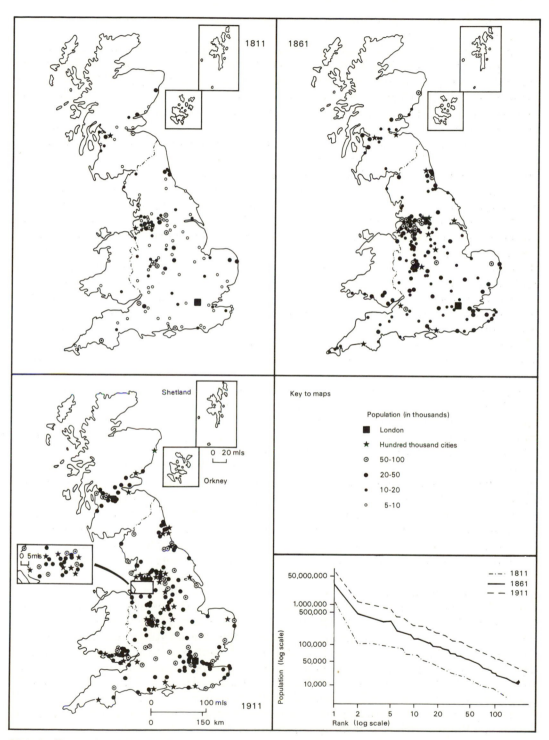

22.1–4   Town size, 1811–1911

always misses the isolated and specialist places like railway Swindon and fishport Grimsby.

This growth involved very few 'new towns' after the fashion of the planted medieval boroughs or the American frontier. In 1800, Britain was already well endowed with population centres and growth involved the development of those nuclei. The few new towns were related either to resource exploitation or were leisure resorts. Their nature revealed a great deal of the processes shaping nineteenth-century urbanization. Middlesbrough was laid out in 1830 as an extension of the better-known Stockton and Darlington coal-exporting system. By 1853 (22.8), the original town layout had been filled. It was a regular grid with a central market square that would have suited an Irish plantation town. This gave Middlesbrough a very different focus from the medieval borough centres like High Street, Glasgow or Briggate in Leeds. During the 1840s, Middlesbrough acquired a dock, an Improvement Act, a pottery, shipbuilding and ironfounding industries. There were several elements here characteristic of urban formation: local merchant enterprise, London capital and professional expertise, a local resource base, railways and the compelling efficiency of the grid in allocating land use. By 1891–3 (22.9), the railways had linked Cleveland iron ore to Durham coal, and Middlesbrough, 'this infant Hercules, this youngest child of England's enterprise' (Gladstone) experienced explosive growth from 7431 people in 1851 to 75,532 in 1891. The resulting shape of the town revealed further aspects of British urban development. The logic of the grid despite several attempts to reassert itself had been disrupted by the railway and by the property boundaries of the old agricultural economy. Working-class housing had spread west along the tracks. More spacious villa housing spread south towards a newly donated park. The town centre had jumped the tracks and reformed around new municipal buildings. The placing of the new industrial iron and steel establishments was most significant. The shepherd's hut and Newport Marsh were replaced by the ironmasters district, whilst to the east was a series of works, each tied to its own settlement of working-class housing, such as Cargo Fleet (1852) and Ormesby (1854). The major industrial sites were placed around the periphery of the existing urban centre, seeking cheap land and access to the new transport system of rail and dredged river, little different in their logic from the modern industrial estate by the motorway. There was little tidy logic in British urban development. Britain lacked strong government direction in land allocation and rarely had a strong local unified landownership, so that the untidy logic of previous land boundaries, access to transport, and competition for desirable environments by a population with dramatically unequal incomes, produced these varied patterns even on the *tabula rasa* of the south bank of the Tees.

Eastbourne, on the south coast, was created for a very different purpose with a much more unified landownership structure dominated by the Dukes of Devonshire and the Gilbert family (Cannadine, 1980). By 1898 (22.7), development showed several characteristic features. To the west of the central business and leisure district was an area controlled by the Duke. His control plus its pleasant position ensured a uniform development of high-quality villa houses standing in large gardens. In outline it was very like the high-status middle-class housing of the large towns. Gilbert's land, slightly less desirable but equally well controlled, was laid out in well-appointed terraces. Mixed ownership south of the railway produced meaner working-class terraces. Strict control and oligopolistic ownership patterns accentuated the social differentiation of the town,

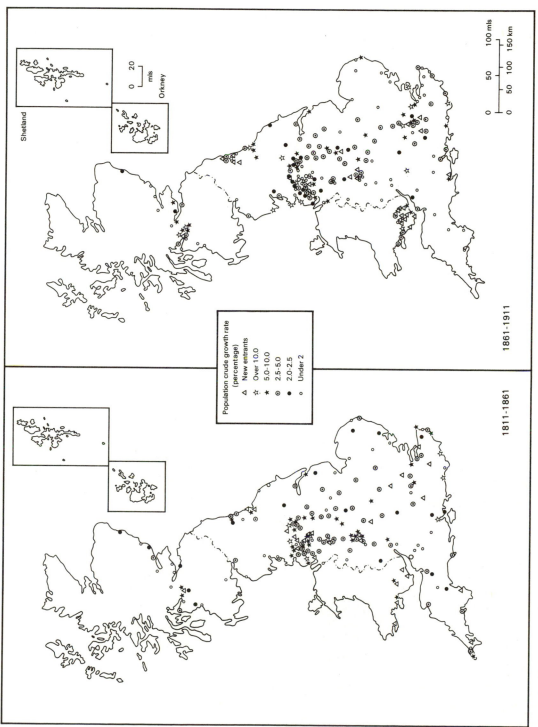

Shetland

Orkney

0 20
mls

Population crude growth rate
(percentage)
△ New entrants
☆ Over 10.0
★ 5.0-10.0
⊙ 2.5-5.0
● 2.0-2.5
○ Under 2

1811-1861

1861-1911

0   50   100 mls
0   50   100 150 km

22.5–6 Urban growth rates, 1811–61 and 1861–1911

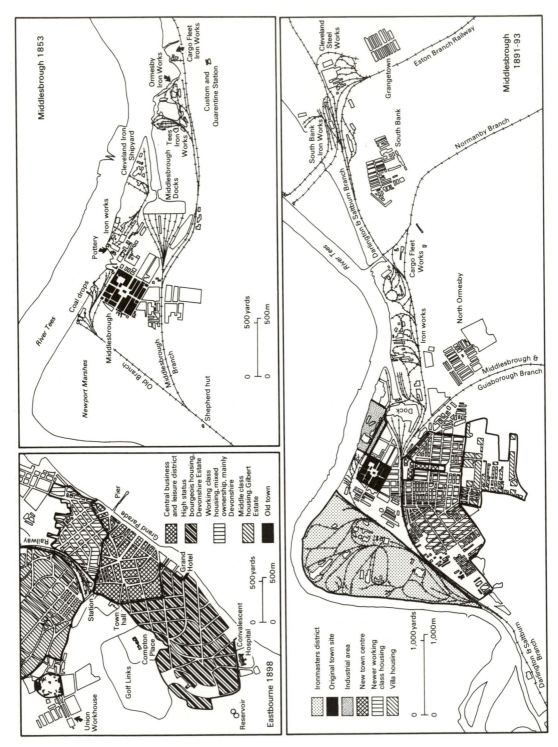

Middlesbrough 1853

River Tees
Newport Marshes
Cleveland Iron Shipyard
Pottery
Iron works
Coal drops
Middlesbrough Docks
Middlesbrough Tees Iron Works
Ormesby Iron Works
Cargo Fleet Iron Works
Custom and Quarentine Station
Middlesbrough Branch
Old Branch
Shepherd hut

500 yards
500m
0

Middlesbrough 1891-93

Cleveland Steel Works
Grangetown
Eston Branch Railway
South Bank
South Bank Iron Works
Normanby Branch
River Tees
Darlington & Saltburn Branch
Cargo Fleet Works
Iron works
North Ormesby
Middlesbrough & Guisborough Branch
Dock
Darlington & Saltburn Branch

Ironmasters district
Original town site
Industrial area
New town centre
Newer working class housing
Villa housing

1,000 yards
1,000m
0

Eastbourne 1898

Railway
Pier
Grand Parade
Station
Town hall
Compton Place
Grand Hotel
Convalescent Hospital
Golf Links
Union Workhouse
Reservoir

Central business and leisure district
High status bourgeois housing, Devonshire Estate
Working class housing, mixed ownership, mainly Devonshire
Middle class housing, Gilbert Estate
Old town

500 yards
500m
0

22.7–9 Social and economic structure of new towns

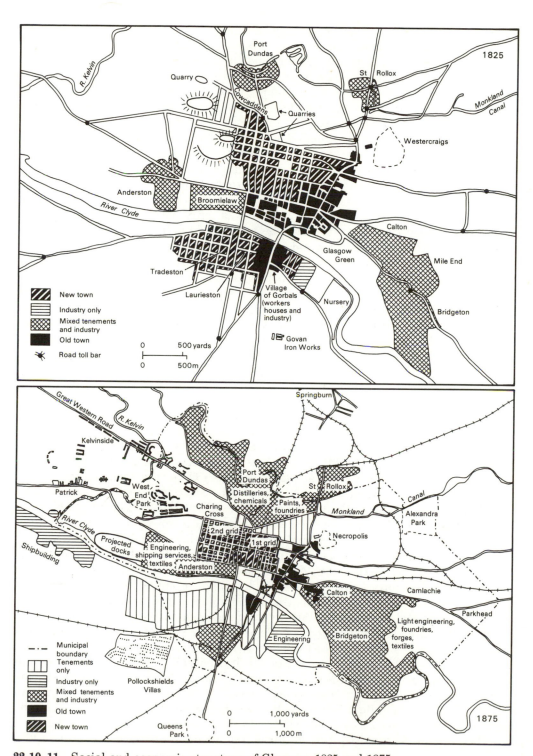

22.10–11 Social and economic structure of Glasgow, 1825 and 1875

whilst the area under fragmented ownership became the less socially prestigous area. Such an area was in any case necessary to provide labour for the construction and service sectors of the local economy. Gilbert and Devonshire between them prevented the creation of boarding houses like those in far away Blackpool.

Most British cities grew by accretion around an existing urban core. Glasgow must stand as an example of this process. By 1800, the old centre based upon Glasgow Cross and High Street had become increasingly unpleasant under the impact of population pressure and industrial activity (Checkland, 1964). The wealthy and middle classes migrated west to the gridiron terraces of the 'new town', followed by the commercial centre. Beyond Glasgow Green, the east end became a mixed area of working-class tenements, textile factories and transport facilities (22.10). There was no sense of Chicago-school concentric rings in these cities. The traditional British analysis of east end, west end and central business district was only partially adequate. Like Middlesbrough, Glasgow was surrounded by a ring of partially discrete industrial settlements: Port Dundas (distilleries and foundries), Cowcaddens (quarries), St Rollox (chemicals), Anderston (handloom weaving) and Govan (iron and coal). Again, these were tied to transport (the canal) and existing patterns of activity. As the area of Glasgow expanded, the wealthy and middle classes maintained the quality of their environment in that northwest sector, in part by their feu charters, in part by out-bidding other land users, and in part because the railways and working-class tenement builders followed the momentum of the earlier period. The developers' planning could rarely hold the logic of industrial development. The select development of Laurieston to the west of the village of Gorbals was rapidly submerged by the external diseconomies of the nearby ironworks and the Govan coal company's railway. By the mid-1870s (22.11), industry looking for large cheap sites had created a new ring of industrial satellites. The North British Railway Company was at Springburn whilst Singers sewing machines and the iron shipbuilders chose Clydebank down the river.

Working-class housing was created in three basic ways – subdivision of existing properties, infilling and new building. All three processes were guided by the profit motive in a free market. The first example comes from Oxford, a town central to the service and agricultural economy. Map 22.12 shows the St Ebbe's area of the city in 1817 (Morris, 1971). It was a frequently flooded meadow, still marked by the defences of the civil wars. Around its fringes were market gardens and a tan-yard, characteristic of early nineteenth-century peri-urban land use. Between 1811 and 1821 Oxford's population increased by over 20 per cent. Building began in 1820 after the auction of Alderman Bricknell's estate (B and C, 22.14). In the next decade the area was laid out in streets and parcelled into small house-sized lots. The boundaries of the lots and direction of the streets reflected the old meadow boundaries and the watercourses. The lots were purchased and built upon in ones and twos by local men of small capital, many of them in the building trade: a mason, a college servant, a tailor, a bookbinder, and a carpenter who became a publican. In 1840, nearly half the houses were owned by those who lived in St Ebbe's. There were two major results of dividing this space into lots and building the houses in twos and threes. Visually, there was infinite variety upon a theme (22.13). Each house or group of houses varied in small details, in the choice of brick or in the height or absence of cellars below the houses. This gave considerable relief to the potential dullness of some of the poorest streets in Oxford. The other result was less happy. When Henry Wentworth Acland, Lee's Reader in Anatomy at the University,

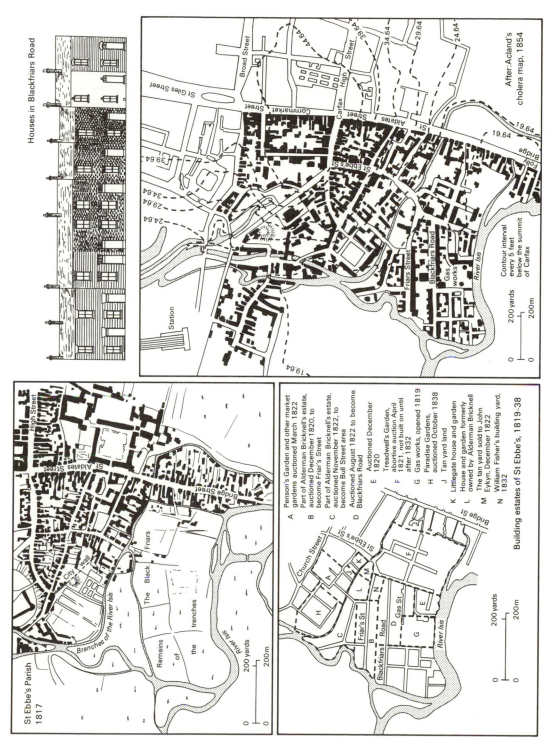

22.12–15   Building development of St Ebbe's parish, Oxford, 1817–54

wrote about the cholera epidemic of 1854 he produced a map (22.15) which directed special attention to the recently built houses of St Ebbe's. The black dots and squares which represented the deaths in that and previous epidemics were concentrated in those new houses. What had happened was that each developer had behaved with impeccable logic within the boundaries of his own plot. A well was sunk for water supply. A soak hole was dug for human waste. However, St Ebbe's was a low-lying, waterlogged area liable to flooding. That was why the cost of the land was low enough to make it profitable for low-income housing. Now cholera is principally a waterborne disease, so that the environment of St Ebbe's was ideal for carrying infection from soak hole to water supply to new victim. Rational decisions taken by individuals under the constraint of the free market added up to the collective disaster recorded on Acland's map.

The example of infilling comes from the textile metropolis of Leeds. William Lupton had been in financial difficulties before his death but left his widow with land at the north end of Leeds. The long narrow closes of the enclosed fields of the manor of Leeds were already occupied by a woollen mill and its reservoir and with the house and outbuildings of a gentleman merchant. As often happens, incidents in the family history of individuals, in this case Mrs Lupton's widowhood, triggered the building at local level. It was only at the aggregate level that the overall pattern of building cycles appeared. In the early 1830s the property was divided into lots and advertised for sale (22.16). Again a selection of builders and tradesmen were the developers. The result was an untidy variety of property, despite the attempt of Mrs Lupton and her advisers to impose order by laying out Merrion and Belgrave Streets. Map 22.17 shows the variety of property, back-to-back and through houses, an independent chapel and the increasingly cramped merchant houses.

An important shift in the spatial organization of this housing has been pointed out by Martin Daunton (1983). It is well illustrated by the maps of working-class housing in Nottingham drawn from the 1844–5 Health of Towns commission (22.18–19). Whilst towns have always had streets, the poorer housing was grouped in a disorderly manner in courts and fold yards. Sometime in the 1820s, often under the influence of local by-laws, town after town began building its working-class housing in streets – easier to police, to light, to drain and to provide with services.

By the 1900s, successive building cycles had accumulated hundreds and thousands of working-class dwellings in British cities. Industrialization did not bring uniformity. We do not yet know enough about regional building styles to map them with precision; although the broad outline of tenements in Scotland, back-to-backs in the north and Midlands are clear, details such as the Tyneside flats and Lancashire yarded houses are not. The information gathered for the Board of Trade cost of living enquiries in 1908 showed the results of regional variation (22.20 and 21). Scotland was a high cost, overcrowded area. The north was overcrowded and low cost, whilst London was a high cost focus.

British towns involved large accumulations of fixed capital in services and infrastructure. By 1850, nearly £12 million at historic cost had been invested in the fixed capital of the gas industry (Falkus, 1967). Most towns over 2500 in population were supplied. It was used mainly for lighting streets and business property. The major accumulations had been incorporated by Act of Parliament (22.22). The distributive pattern of gas prices (22.23) was in part a regional one reflecting access to suitable coalfields and to the cheap transport costs of the coastal route, and, in part a reflection of town size, company

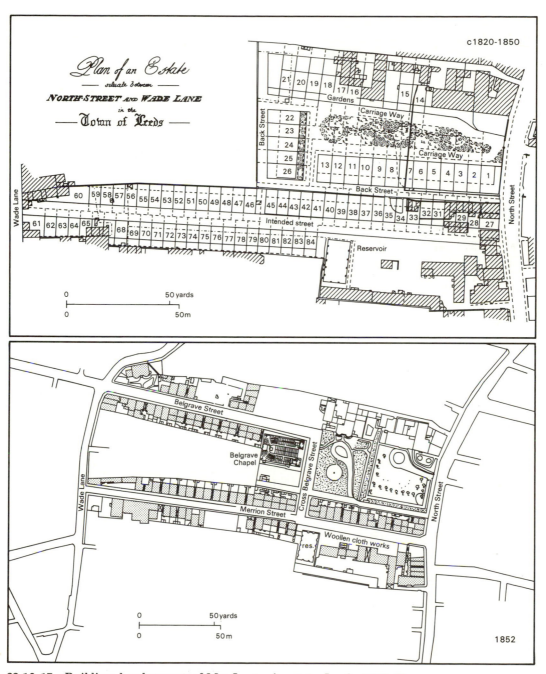

22.16–17 Building development of Mrs Lupton's estate, Leeds, 1820–52

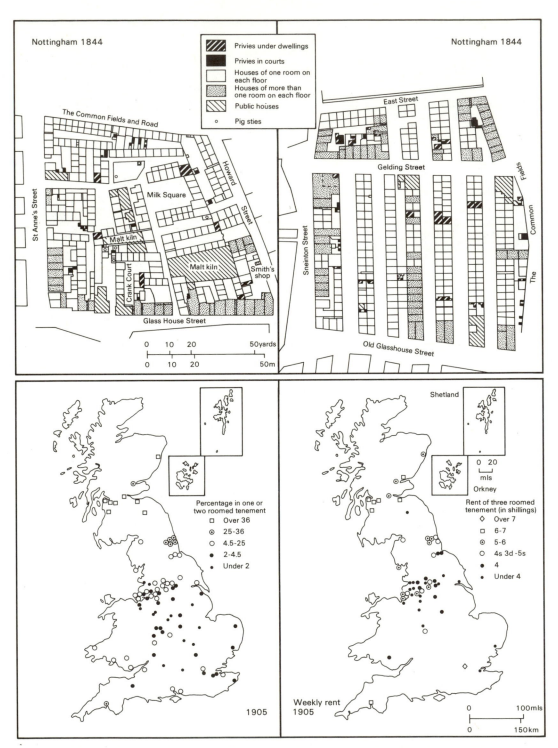

22.18–21 Nottingham housing, 1844; British housing accommodation and rents, 1905

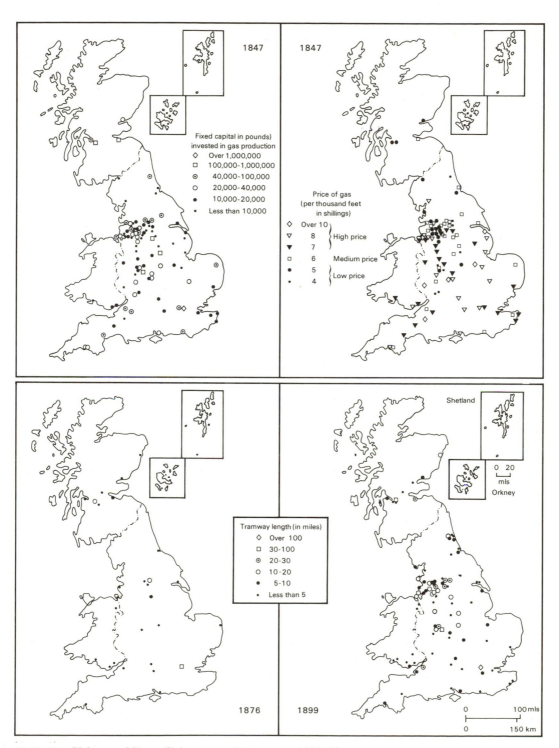

22.22–25   Urban public utilities: gas, 1847; trams, 1876–99

age and management efficiency. Tramways also needed parliamentary approval. The first wave (22.24) showed that London, Scotland and Yorkshire were areas of innovation. By 1899 (22.25), the overall pattern of large urban centres had begun to assert itself.

By the third quarter of the nineteenth century, the larger urban centres confronted the need for major inner-city road investment. In Edinburgh, Chambers Street was driven through crowded eighteenth-century squares and wynds under the 1867 City Improvement Act, a result of the constant nagging of the city's first Medical Officer of Health, Dr Littlejohn, and the inspiration of the patriarchal Lord Provost, William Chambers (22.26). The street had a double purpose. It was an east–west link between two major north–south routes and avoided the congested High Street. A warren of courts and squares was changed into a major institutional street containing the University, the Watt Institution, the Church of Scotland Normal School and the Museum of Science and Art. A similar move in Birmingham produced Corporation Street, a major shopping centre. There were several such ventures in London, like the Embankment and Queen Victoria Street. These streets drove space, light and air into crowded areas. They were a spatial embodiment of miasmatic theory (stale air and smells cause disease). The density of urban activity was such that these developments need to be seen in section as well as plan. When the recently created London County Council built Kingsway from Holborn to the Strand under an act of 1899 (22.27), the roadway hid a variety of subways and vaults to impose order upon the service and transport needs of the new street.

These services and new forms of transport were themselves a means of transforming the shape and use of cities. The British pattern was that of Edinburgh in which the tram and train followed the builder and by doing so stimulated further advance within walking distance of the terminus. The 1895 map (22.28) showed the fingers of new building in the suburbs of Morningside and Newington. The stations of the surburban railway provided another link to the centre. The London 1905 map (22.29) showed the massive rent gradient that was developing in the expanding cities. This meant that those with adequate income and regular work could use the new forms of transport to make dramatic improvements in the space and quality of their environment. The electric trams from Greenwich to Waterloo or from Enfield to the City cost a shilling a week at the early-morning workmen's fare.

During the nineteenth century the social characteristics of the population began to assume distinct spatial patterns, although historians are not as sure as they once were about the exact nature of these distributions. David Ward's (1980) analysis of status in Leeds showed that simple class-segregation was an inadequate account. There were areas of high status especially in the west end and west suburbs, but most other areas contained significant proportions of high-status residents (22.30). More detailed analysis of neighbourhoods and their patterns is necessary before surer statements can be made. Nor did distributions of non-English immigrants in the big cities fit the ghetto pattern in all cases. In Liverpool in 1871 (22.31), the Irish in the north and centre best fitted the ghetto description, but the Welsh of Everton (northeast) and Toxteth (south) were an ethnic community without the pressures of poverty and low status (Pooley, 1977). The relationship of gender to space is perhaps the most enigmatic to historians. Leeds in 1841 (20.32) shows that higher proportions of women were found in high-status areas. These were areas full of female servants but also with many middle-class widows and spinsters. There was also a slight tendency for the female proportion to rise in the lowest-status areas, perhaps reflecting the low-paid female labour in textiles.

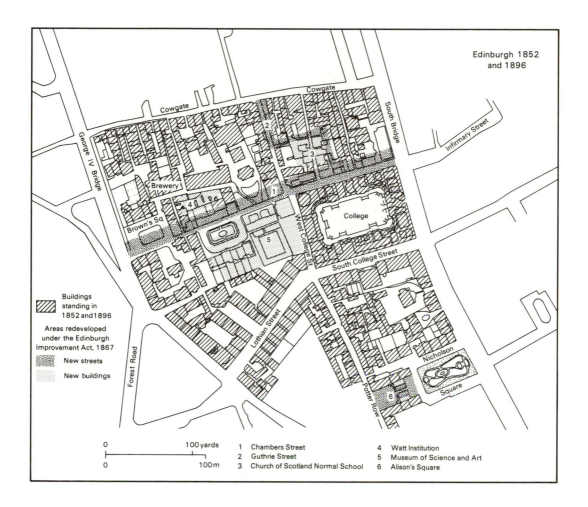

Edinburgh 1852
and 1896

Buildings
standing in
1852 and 1896

Areas redeveloped
under the Edinburgh
Improvement Act, 1867

New streets

New buildings

| 0 | 100 yards |
| 0 | 100 m |

1	Chambers Street	4	Watt Institution
2	Guthrie Street	5	Museum of Science and Art
3	Church of Scotland Normal School	6	Alison's Square

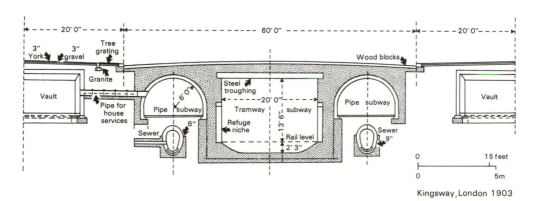

Kingsway, London 1903

22.26–27 Urban improvements: Edinburgh and London

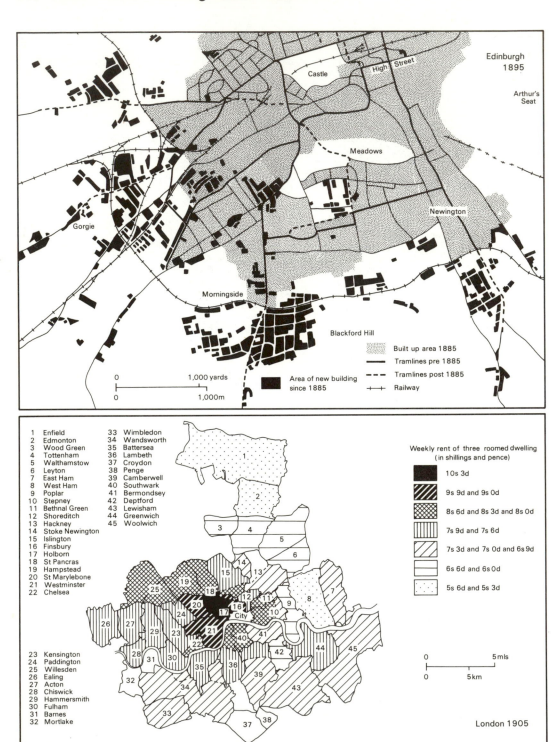

Tramways and building in Edinburgh, 1885–95; rents in London, 1905

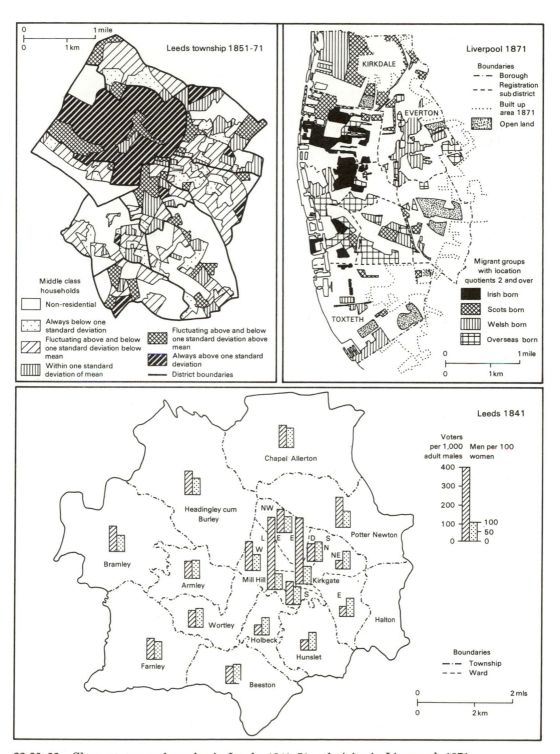

22.30–32   Class, status and gender in Leeds, 1841–71; ethnicity in Liverpool, 1871

23 **Retail patterns**

Gareth Shaw

The industrial revolution witnessed considerable changes in the distributive system, particularly with regard to the form and pattern of retail facilities. Increasing levels of economic and urban growth led to the concentration of purchasing power which favoured the widespread development of retailers operating from fixed shops. In the late eighteenth century, owing to poor levels of inter-urban transport which localized the production of consumer goods, many of the shops were run by craftsmen retailers. By the 1850s and 60s the distribution system had fragmented into more clearly defined groups of producers, wholesalers and specialist retailers in response both to changes in demand and developments in supply technologies. The full force of these changes occurred in the latter part of the century, with the growth of large-scale retail organizations and department stores.

By 1830 shops had already become important supply outlets throughout the country (Alexander, 1970). In most towns the highest rates of shop growth occurred from the end of the eighteenth century to around 1820. Indeed, in many cases shop numbers increased faster than population until the second half of the nineteenth century. In ten sample towns average rates of population per shop fell sharply from 136.3 in 1801 to 56.0 by 1881.

It is extremely difficult to measure the pattern of shop growth at a national level owing to data limitations and the task of abstracting information from a great number of local directories. Furthermore, many of the most important patterns of change happened at the intra-urban level. Thus shop growth was conditioned by the rapid rates of population suburbanization found in nineteenth-century cities which created new areas of consumer demand away from established central retail facilities. Such patterns of retail change are illustrated in Map 23.1, which shows an example of the evolution of major shopping streets. By the 1820s shops were already locating, in relatively large numbers, outside the central area. However, even by 1851, limited levels of mobility restricted the degree of shop dispersal. At this date shopping facilities occurred in a fairly clustered pattern, concentrating in the centres of the main areas of working-class demand. By the 1880s the picture had changed. The growth of numerous linear shopping streets reflected the greater levels of suburbanization made possible by urban tramways.

Increased consumer demand induced changes in other forms of retailing, particularly markets and itinerant traders. The growing importance of markets, especially as major places of food retailing, led both to organizational changes and physical restructuring. The separation of retail and wholesale markets had been initiated before 1800, and the nineteenth century witnessed further important changes. Thus a great many new and expanded markets were created after 1800 (23.3), which represented significant developments in a wide range of settlements. Such changes were not confined entirely to large industrial centres but occurred throughout the urban hierarchy. The scale of development is illustrated by the fact that between 1800 and 1890 sixty-four towns (excluding London) had obtained Parliament's permission to change their markets, and as Map 23.3 shows, the majority of such changes concerned the creation of new markets. In addition substantial developments occurred within London, with fourteen new markets being opened during the nineteenth century.

Urbanization also gave a boost to certain types of itinerant traders, whose numbers increased from 14,662 in 1841 to 58,939 by 1891. Some caution is needed, however, in using these census-based figures since the transient nature of this occupation probably precluded accurate enumeration. Furthermore, it seems likely that numbers fluctuated

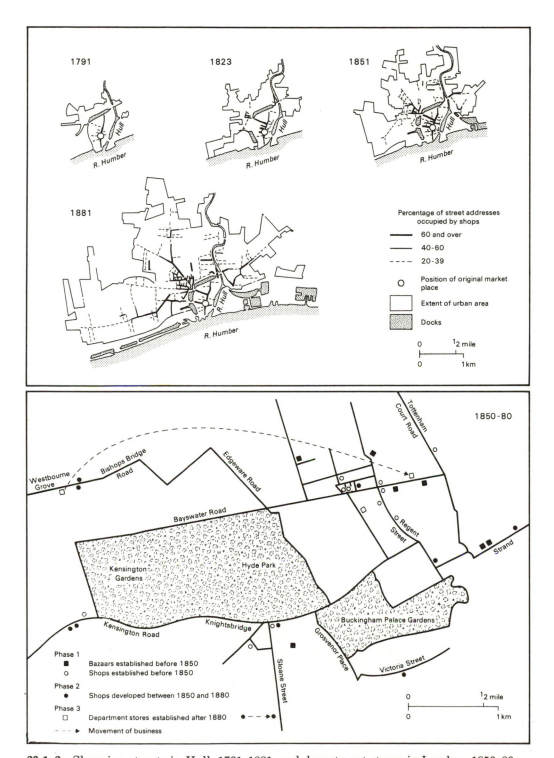

23.1–2 Shopping streets in Hull, 1791–1881, and department stores in London, 1850–80

in response to general levels of employment, with some unemployed workers moving into petty trading. The growth of itinerant traders was linked to urban development (23.4). Heavily urbanized counties, such as Lancashire, the West Riding and the central lowlands of Scotland, had the highest proportions of traders. In contrast most of the predominantly agricultural counties, especially in Wales, northern Scotland, and southwest England, possessed very low levels of itinerant traders. This situation was probably further intensified by an absolute decline in the more established country trade, possibly through a combination of rural depopulation and competition from shops. Most of the rural hawkers operated on a larger scale than their urban counterparts and required a licence. The number of licences issued therefore gives a measure of the rural trade, which reached a peak in England and Wales of 7479 traders by 1830 but had declined to 5742 by 1843.

In the second half of the nineteenth century, continued urban growth and the expansion of consumer demand presented opportunities for the development of large-scale retailers such as multiple retail firms and department stores (Jeffreys, 1954). Consumer co-operatives represented a third major type of retail organization, based on political as well as economic factors. Unlike the multiples the co-operative stores had many of their initial successes in the smaller industrial settlements. The earliest growth up to 1870 was in the mining and textile areas of northern England, followed by the east Midlands and later by Scotland (23.5). In contrast, developments in London and southern England came not only later, but were much smaller in number. Nevertheless, at a national level co-operative societies dominated grocery retailing, as their membership increased from 500,000 in 1880 to 1,700,000 by 1900.

Multiple retailers grew most rapidly after 1880, and by 1900 there were 257 organizations controlling about 11,650 shops. Lack of data makes it impossible to discuss patterns of growth, although limited evidence suggests that early developments focused on large cities. Estimates for the grocery trades allow some of the results of differential rates of regional growth to be illustrated (23.6). Once again relatively large numbers are found in the industrial north and Scotland. However, unlike the co-operatives, multiple retailers were also biased towards London and the home counties, and to a lesser extent the Midlands.

While multiple retail firms gained scale economies by operating large numbers of shops, department stores attempted to obtain advantages by increasing the range of articles sold in one store (Shaw and Wild, 1979). This resulted in larger shops, initially brought about through the amalgamation of adjoining premises. The growth of department stores is difficult to trace nationally, but their growth in London's West End indicates some of the evolutionary stages. As Map 23.2 shows, three main phases may be recognized. The first centred around the construction of fashionable bazaars, buildings where retailers could rent a stall, during the 1840s. At the same time some retailers were starting to expand their shops by buying adjoining premises. Such amalgamation processes increased rapidly in the second phase between 1850 and 1880, in an attempt to increase store size. The final phase of growth occurred after 1880, with the construction of purpose-built department stores, as typified by the opening of Selfridges in 1909.

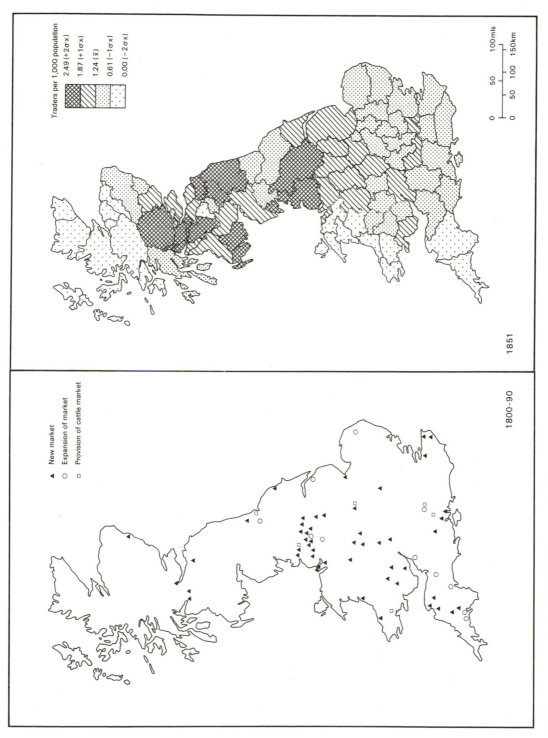

23.3–4 New markets, 1800–90; itinerant traders, 1851

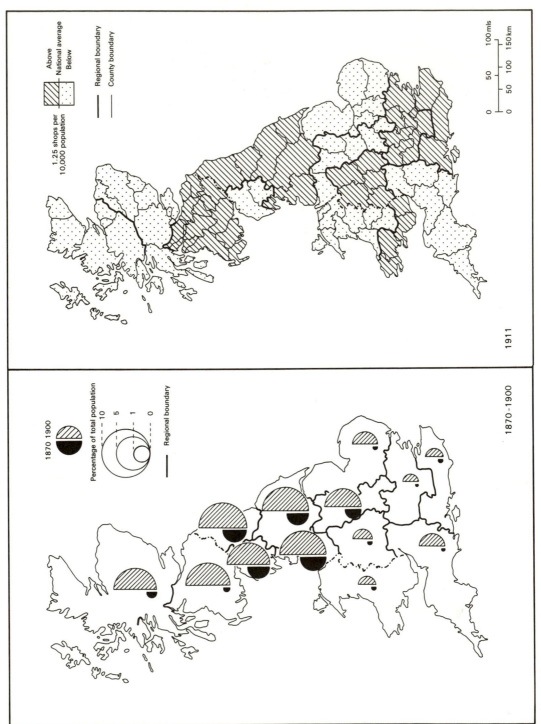

23.5–6 Co-operative societies, 1870–1900; multiple grocery retailers, 1911.

Labour protest 1780–1850

Andrew Charlesworth

The most frequently staged collective protest by industrial workers before 1802 was the food riot. The years 1794–6 were the climacteric of such protests in the eighteenth century (24.1). Food riots sprang from a consumer consciousness set within the world picture of the small producer and the artisan, with its control and regulation of the process of exchange and the labour process, which underpinned much of the labour protest between 1750 and 1850 (Charlesworth, 1983).

Labour disputes, which were also widespread geographically in the last two decades of the eighteenth century, represented similar reiterated collective pressures to take action over wages, working hours, entry to the trade and the use of machinery (24.2). The London bias in the spatial distribution is more a matter of the source material than the propensity of the capital's artisans to disputation.

Food issues became less important in the minds of the main body of industrial workers as the dynamic of industrial capitalism became revealed in its cyclical downturns of unemployment and wage cutting and pulses of technological innovation. In this transitional period, however, industrial workers were quite prepared to adopt whatever seemed the most appropriate form of collective action, be it trade unionism, petitioning Parliament, strikes, machine breaking, food riots or even plans for insurrection, to defend either their craft status or their livelihood.

This is reflected in the geography of the three discrete series of disturbances of 1811–12 that have come to be known as Luddism (24.3). In southeast Lancashire the handloom weavers, having unsuccessfully petitioned Parliament, set up secret committees and took oaths in planning attacks on the new powerloom mills. Divided councils amongst the committees over these tactics and the invulnerability of such large factories to attacks by crowds restricted the number of overt protests and turned the workers to take action in food riots and then to arming themselves. In the West Riding of Yorkshire the croppers' early success against the shearing frames in small workshops was similarly tempered when they turned their attention to larger concerns. Again the 'physical force' advocates held the day for a time. Only in Nottinghamshire was a highly organized and disciplined campaign of attacks on stocking frames by framework knitters over the breakdown of custom in trade sustained over a long period. Having achieved some of their desired ends and faced by determined repression by the authorities, the knitters switched to constitutional agitation. In this they were unsuccessful. Only the disturbances that were out in the open can be charted with any degree of locational precision.

The remoteness of Luddism from London, the centre of political power, meant that the workers' moves towards insurrection were bound to be shortlived. The geography of the agrarian Luddism in 1830 (24.4) would superficially suggest that the farmworkers had more chance of a 'Revolutionary Movement'. They had, however, no tradition of anti-state labour solidarity to mount such a challenge. The pauperized mass of labour in the countryside set limits on their degree of collective action. The severity of such limits has, however, been exaggerated, as the events of 1830 indicate. The show of strength frightened the village élites who henceforth determined to replace the paternalist code of rural society with the harsh rules of capitalism and the New Poor Law. Farmworkers' protests went underground in the 1840s.

The failure of attempts at general unions, the political breach with the middle classes over the 1832 Reform Bill and the anti-worker stance of the reformed Whig administration convinced workers that to protect themselves against the deepening crises of

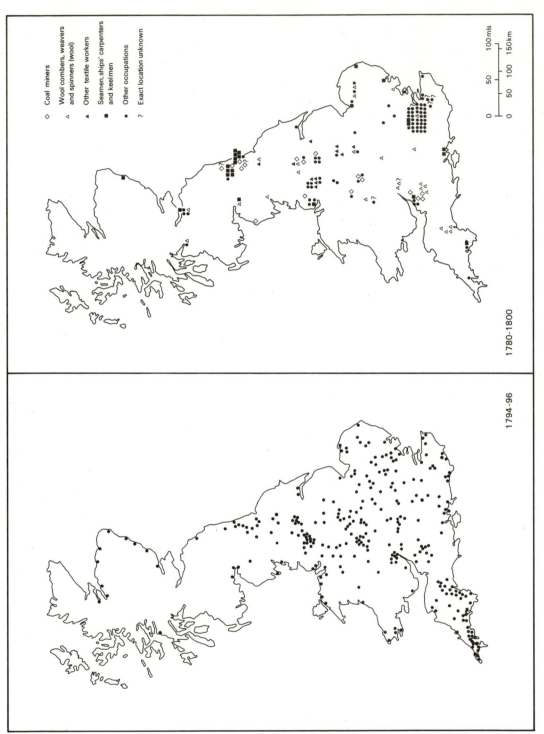

24.1–2   Food riots, 1794–6; labour disputes, 1780–1800

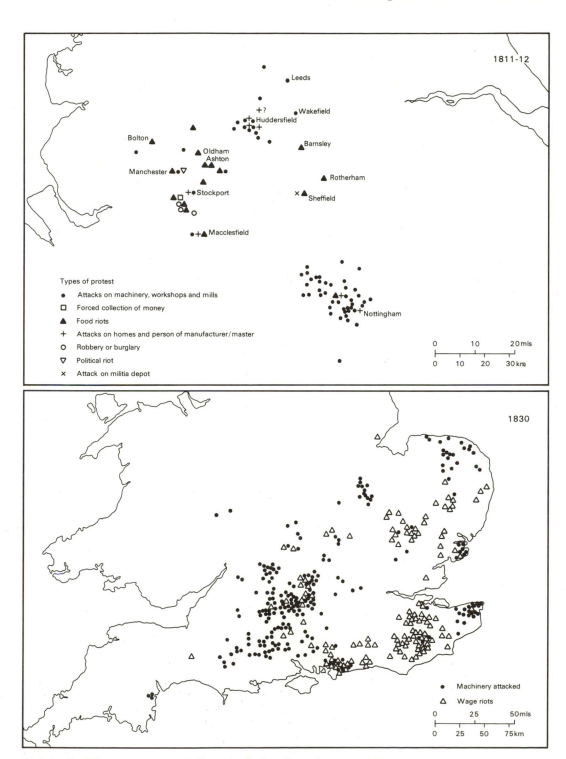

Types of protest

- ● Attacks on machinery, workshops and mills
- □ Forced collection of money
- ▲ Food riots
- + Attacks on homes and person of manufacturer/master
- ○ Robbery or burglary
- ▽ Political riot
- × Attack on militia depot

1811-12

Leeds

Wakefield

+ ?
Huddersfield
+

Bolton

Barnsley

Oldham
Ashton

Manchester

Rotherham

Stockport

× Sheffield

Macclesfield

Nottingham

| 0 | 10 | 20 mls |
| 0 | 10 | 20 | 30 km |

1830

- ● Machinery attacked
- △ Wage riots

| 0 | 25 | 50 mls |
| 0 | 25 | 50 | 75 km |

24.3–4   Luddism, 1811–12, and Captain Swing disturbances, 1830

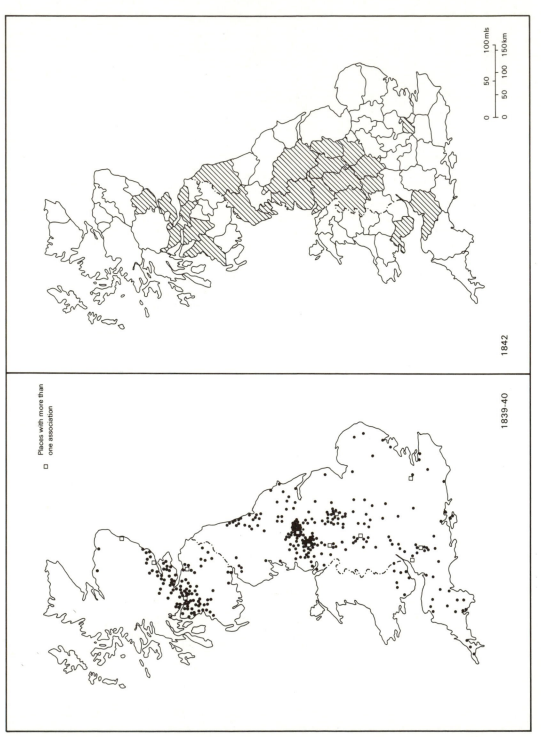

□ Places with more than
   one association

1839-40

1842

24.5-6 Chartist associations, 1839–40; counties affected by the General Strike of 1842 industrial capitalism they had to gain power in the national political arena by their own efforts. Thus Chartism was born. Its focus on political reform and its artisan world picture appealed to handworker, craftsman and factory worker alike. This is reflected in the complex topography of Chartism (24.5). As with Luddism, Chartism had its physical force' (insurrection) and 'moral force' (petitioning) wings, the former finding its greatest support amongst the handworkers, save in Scotland, and the latter its support amongst the urban craftsmen. Direct links between labour unrest and the demands for political reform by Chartists have been questioned, but there seems little doubt that the spatial extent of the 1842 'General Strike' owed as much to the political excitement of that year as to the economic conditions of the different industries affected (24.6).

The unity of the Chartist movement was also based on the belief by workers that the future existence of capitalism still hung in the balance. When faced with the inevitability of the new factory system of production and hence with the redundancy of the ideology of the artisan outlook, workers abandoned the Chartist cause and were forced to learn a new set of 'rules of the game' in which 'collective bargaining by riot' had no part to play.

25   Unionization
Humphrey Southall

At its simplest, 'unionization' is a purely statistical matter: the proportion of the employed population, nationally or locally, belonging to a trade union. Unfortunately, comprehensive statistics for the union movement date only from the 1890s, while the history of early unionism is primarily one of myriad small and local societies. Their ambiguous legal position, particularly before 1825, further impedes systematic documentation. This section therefore seeks to identify the principal foci of early unionism and outlines its organizational development. Evidence of the distribution of early trade unions came from the registrations of friendly societies with local magistrates, following an Act of 1793; another major source is the rulebooks of individual unions.

Before the welfare state, 'friendly' provisions such as sick and funeral benefits were an important aspect of unionism, and many major unions initially registered as friendly societies; the resulting records were centralized by the Registrar of Friendly Societies. Map 25.1 shows all societies in England and Wales first registered before 1810 whose name indicated an occupational basis, excluding state employees, clerks and professional groups. Although scarcely comprehensive, the resultant pattern corresponds well with other more impressionistic evidence. The named trades are those containing five or more societies in a given county, in descending order of frequency. The inset relates numbers of these trade societies in each county to the 1801 population. Three main centres of unionism emerge: London, containing a third of all societies, in a wide range of occupations, particularly artisan and riverside trades; the Lancashire cotton industry; and the northeastern coal trade, meaning sailors, keelmen and shipwrights

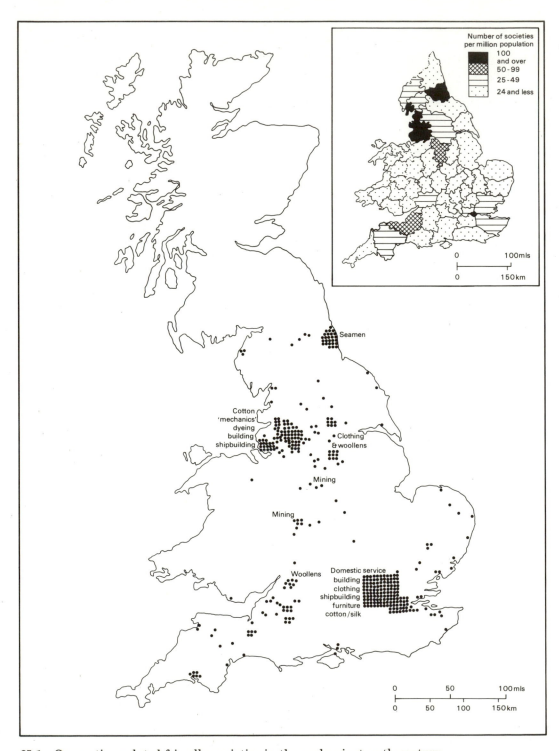

25.1   Occupation-related friendly societies in the early nineteenth century

25.2–5   Trade friendly society branches in the 1820s

rather than miners. The southwest, with its woollen industry, appears as a secondary concentration, while Liverpool and Bath resembled London, with a range of luxury artisan trades. It is impossible to estimate total membership of these early societies, but 100–500 was probably typical for a single club in a large centre.

The earliest societies were single-branch, meeting at a trade's 'house of call', or pub. National unions developed out of reciprocal arrangements between such bodies, generally appearing first among artisans as a means of assisting 'tramps': travelling members seeking work, carrying cards entitling them to bed and board at each house of call. Maps 25.2–5 show the components of four such bodies in the 1820s, all drawn ultimately from rulebooks; although such national 'unions' must have existed before, they only declared themselves publicly at the time of the repeal of the Combination Acts, and these are the earliest societies for which such maps can be drawn. For the Brushmakers, the circuit required of tramps is also shown; records of the actual movements of individual tramps show that such extended journeys were not uncommon. After 1850, major national unions such as the Amalgamated Engineers and the Carpenters and Joiners developed among artisans, but regional autonomy survived far longer in non-tramping trades. The extreme example of the latter was mining, the National Union of Mineworkers being formed only in 1945 and remaining a confederation of coalfield unions, while groups such as cotton workers were inherently localized.

More systematic evidence is available after 1890. Map 25.6 shows total union membership as a percentage of all employed, based on research by the Webbs. As the divisor includes women and clerical workers, largely non-unionized prior to the quadrupling of total membership between 1892 and 1920, it is possible that a majority of the male working-class was unionized in some areas. Conversely, counties such as Dorset, with only 305 unionists representing 0.38 per cent of the total workforce, were almost untouched. The marked north–south divide in this map resembles that in Map 25.1, and perhaps permits us to speak of union and non-union Britain. Map 25.7 shows the distribution of industrial disputes, based on the earliest labour statistics of the Board of Trade. The similarity between this and patterns of unionization suggests an unsurprising link between unionism and labour militancy, despite the great emphasis on 'respectability' among most late Victorian unions.

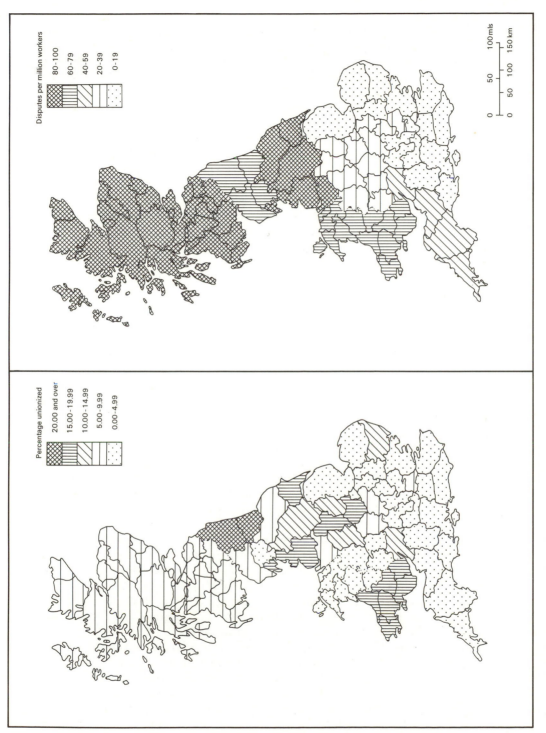

25.6-7 Unionization of employed population and strikes, 1892

26 **Popular institutions**

Martin Purvis

From the late eighteenth century onwards popular institutions developed in a wide variety of fields including education, recreation, social reform and self-help. The new institutions were uneven in their geographical distribution, often being concentrated in urban and industrial areas. The newly influential provinces provided the mass support and the initiative of leadership for many popular institutions.

Friendly societies (26.1–2) were the most numerous nineteenth-century working-class organizations. Through regular contributions members provided against illness, injury and death. Early societies also included trade union and co-operative trading functions, and social activities were important. The maps are a guide more to the distribution of societies and its change over time than to their absolute numbers. Based on parliamentary returns (26.1) they omit societies which were not officially registered, and in Map 26.2 registered societies not making an annual return of their affairs. The 1872 returns were fuller than normal but only 56 per cent of English returns were completed. Lancashire displays consistent strength. Other expanding industrial areas, where relatively high wages allowed workers to afford contributions, increased their share of the national total. The contribution of de-industrializing East Anglia and the West Country decreased. The declining number of Middlesex societies was partly offset by the substitution of fewer large societies for a multitude of small ones. The dearth of Scottish societies in Map 26.2 partly reflects the especial inadequacies of the Scottish registration system (compare the numbers returned here with the earlier survey in Levitt and Smout, 1979). Coverage was less than half the suggested rate for Lancashire (*BPP* 1874, XXIII, Parts I & II). Expansion of affiliated orders (societies with a branch network organized from a central headquarters) was important in friendly society growth. The pie charts show the distribution of members across the English counties in 1876 (as shaded on the main map 26.2) for the two largest orders. Both originated in southeast Lancashire. The Oddfellows were strongest in the industrial north while the Foresters, by allowing flexibility in rates of contribution and benefit, gained members in the lower-wage areas of southern England (Gosden, 1961).

If claims that 700 co-operative societies existed by 1831 were correct then Map 26.3 shows some 55 per cent of this total. Co-operative retailing and production enjoyed temporary popularity around 1830. Often linked to attempts to finance Owenite socialist communities, stores and workshops were also an answer to immediate problems of food supply and employment (Garnett, 1972). Initial publicity came from a Brighton society, provoking local emulation and boosting Owenism amongst London artisans. Co-operation was also strong in the textile districts of Lancashire and the West Riding. The failure of societies after 1832 was less complete in the West Riding than elsewhere. Perhaps as a result, the revival of co-operative trading in the late 1830s and early 1840s, largely Chartist inspired, was strongest in the north (Purvis, 1986). Scotland had an independent history of co-operative trading from the eighteenth century onwards, supplemented by Owenite- and Chartist-inspired activity.

Official registration of co-operative societies was more comprehensive than for friendly societies, and provides a more complete record later in the century (26.4). Growth from the 1840s onwards established the geography of later nineteenth-century co-operation (Cole, 1944). Societies were particularly successful in medium-sized towns in the textile districts, the mining villages of northeast England and the shoe and hosiery districts of the east Midlands. Success was related to the size and economic

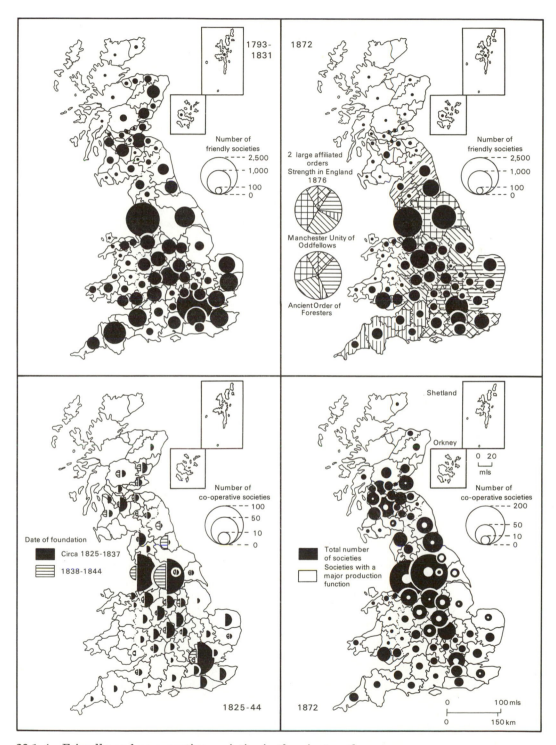

26.1–4   Friendly and co-operative societies in the nineteenth century structure of settlements. Societies were less successful both in large centres, particularly London and Birmingham where small workshop employment predominated, and in low-wage agricultural areas. Most societies were primarily retailers but some funded educational and welfare projects (Attfield, 1981). Investments were made in co-operative production. In the north this included large-scale operations in textiles and engineering as well as craft production in self-governing workshops where co-operative production had first developed (B. Jones, 1894).

Public houses were important foci of working-class life, offering a value system counter to those of religion and self-help. They were centres for exchange of news and goods and the meeting place for clubs and societies (Harrison, 1973). The greatest number of pubs were in industrial and mining areas where disposable working-class income was highest (Levitt and Smout, 1979). However, pub density in relation to population (26.5) was highest in smaller boroughs and counties experiencing little population growth. A reduction in the number of licences granted, especially in Scotland, and a failure of public-house building to keep up with urban population growth led to a long-term decline which was reflected in employment in drink retailing.

Efforts to reduce drinking were particularly supported by the middle and 'respectable' working class. The campaigns moved from temperance to teetotalism in the 1830s, then to the United Kingdom Alliance and the Scottish Permissive Bill Association which campaigned for the local prohibition of alcohol sales (Harrison, 1971; Lambert, 1983; Paton, 1976). In 1872 the Alliance was near the peak of its popular appeal. Support came from urban and manufacturing areas, particularly the textile districts, and other areas of nonconformist strength. It campaigned through petitions and meetings (26.6). Campaigning was limited outside areas of established support – reflected in donations to the Alliance (26.6, inset) – including rural areas where pub density was high (26.5).

The Women's Co-operative Guild (26.7) was formed in 1883, with a separate Scottish Guild in 1889. The importance of the Guild lay not in its size (total English and Welsh membership in 1899 was 12,600) but in the opportunity it gave for initiative by working-class women in a male-dominated age, including campaigns on issues such as suffrage, divorce law reform and health care (Gaffin and Thomas, 1983). By the end of the century the distribution of Guild branches matched the strength of co-operation as a whole. The greatest number of early branches were in southeast England, perhaps because the relatively young co-operative societies of the area were more receptive to the involvement of women than were the long-established northern societies.

Working men's clubs (26.8) had a metropolitan origin, hence the large number in London. The movement was originally based in the 1860s on middle- and upper-class patronage. The village clubs of rural southern England reflect this support. Most clubs were in urban locations, particularly in London, Lancashire and the West Riding, where there were concentrations of higher-wage workers. Clubs were later in developing in northeast England with significant numbers only from the early twentieth century – see Map 26.8 inset for Club and Institute Union affiliated clubs in 1909 (Hall, 1922). This was perhaps because music halls already provided similar facilities. Urban clubs became increasingly independent, developing political activities, professional entertainment and the sale of alcohol (Ashplant, 1981; Shipley, 1971).

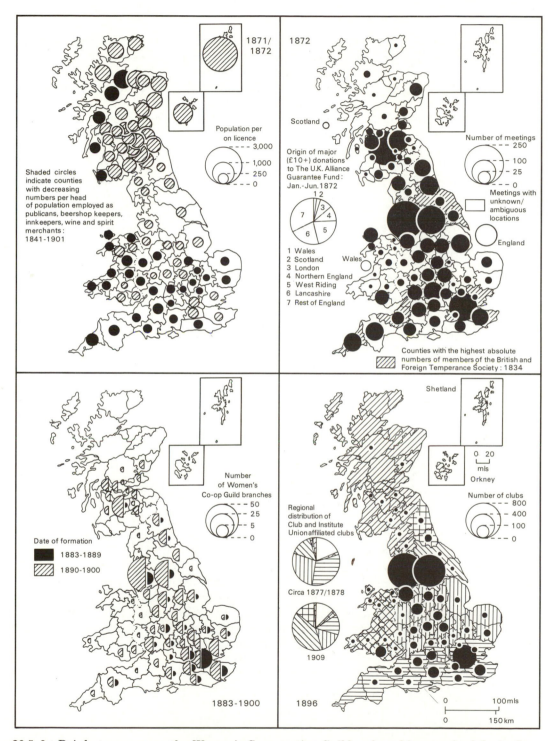

26.5–8   Drink, temperance, the Women's Co-operative Guild and working men's clubs in the nineteenth century

27 Sport
Wray Vamplew

Commercialized sport was one of the economic success stories of late Victorian Britain. In response to rising incomes and Saturday half-days, entrepreneurs and club executives enclosed grounds, invested heavily in facilities, and charged gate-money. Events were organized on a regular basis to offset overhead costs and crowd-drawing tours by overseas teams were encouraged. The development, however, was not uniform. Some areas, notably Yorkshire and Lancashire, were in the vanguard of sports commercialization, but, in contrast, the eastern counties remained relatively untouched. The home counties took to county cricket with great success but spurned the league version; they also pioneered enclosed race-courses but lagged in the adoption of professional soccer. Such football was also slow to intrude into South Wales and never caught hold in the Scottish borders, both areas remaining strongly committed to rugby union.

A striking development in horse-racing (27.1–2) was the growing spatial concentration of meetings simultaneously with a widening distribution of events over the year in order to utilize resources more effectively.

In cricket the public began to prefer inter-county matches to those of the peripatetic professional elevens from the 1870s, though a county championship was not organized by the MCC till 1890. The 1890s saw the development of Saturday afternoon league cricket, but, as Map 27.3 intimates, its semi-professional, competitive nature was anathema to southern amateurs. Map 27.4 indicates that a high degree of playing inequality existed at the first-class level which suggests that profits were not a prime objective of county clubs: indeed several (marked *) had to resort to public appeals in order to survive.

Map 27.5 shows that by 1910 top level rugby union had spread throughout Britain. On route, however, a major schism had occurred within rugby circles as in 1896 a dispute over broken-time payments had led to the emergence of professional rugby league. Although some inroads were made elsewhere, essentially, as is clear from Map 27.6, this remained a Lancashire and Yorkshire sport.

Football became highly commercialized. Map 27.7 shows that large crowds were regularly being attracted to first division matches in England and ground-capacity figures suggest that even more spectators were expected at Scottish games, particularly in Glasgow. The economics of gate-money soccer stimulated clubs to adopt company status, though, as seen in Map 27.8, there was a lag in Scotland and the south of England which stemmed from the slower development of professionalism in those areas. Nationally the occupational composition of succer club shareholders (27.9) differed, with Scotland having a significantly higher involvement of the drink trade and manual workers than England. Map 27.10 shows that commercialization influenced playing success. In England the southern amateurs lost out to the professionals of the north and Midlands; in Scotland Glasgow continued to dominate, but with the professional Celtic and Rangers clubs replacing the still-amateur Queen's Park.

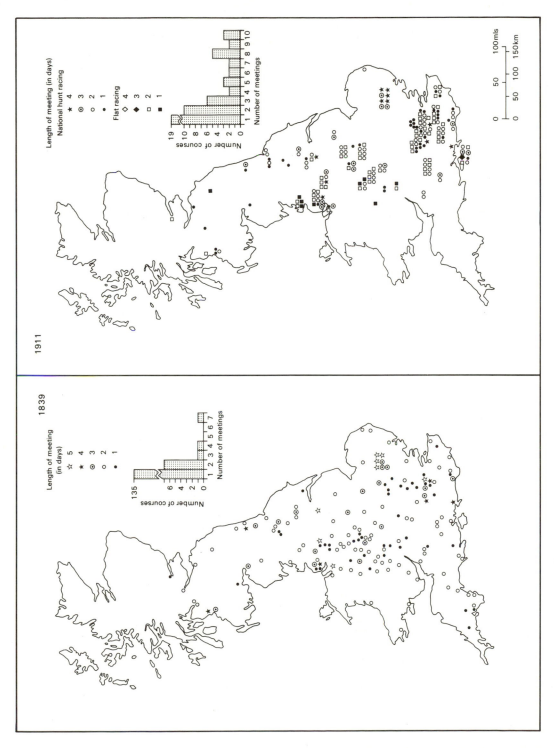

27.1–2   Horse racing, 1839 and 1911

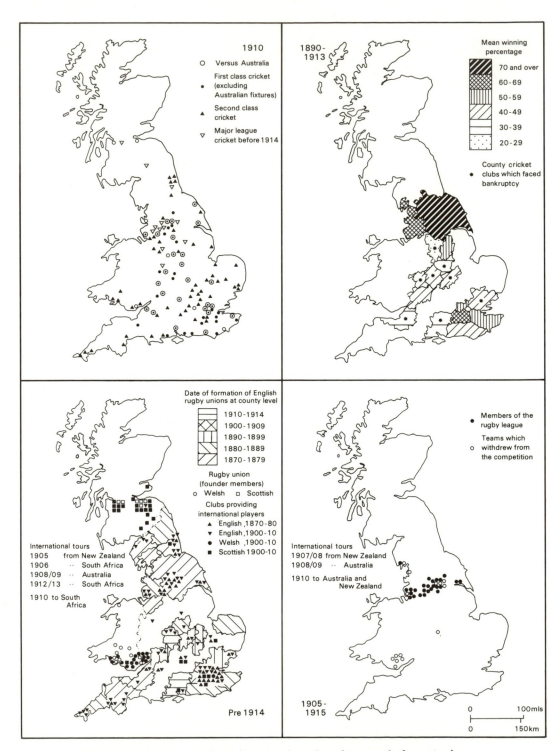

27.3–6 Cricket and rugby in the late nineteenth and early twentieth centuries

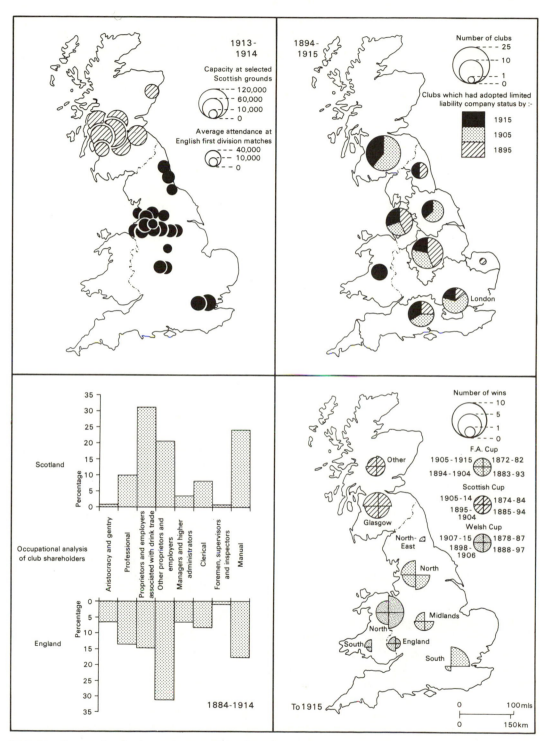

27.7–10 Soccer in the late nineteenth and early twentieth centuries

28 **Languages and dialects**
John Langton

Eight languages were in majority use in different parts of Britain in the early seventeenth century (Ravenstein, 1879): Norn, derived from Norwegian, in Shetland, Orkney and Caithness; French in the Channel Islands; the celtic languages of Cornish, Welsh, Manx and Scottish Gaelic; Scots, which was on most linguistic criteria as different from English as, say, Portuguese from Spanish, was used in southern, central and eastern Scotland, and English in most of England, parts of Wales and metropolitan Scotland. The codification of Standard English was well under way. It was used in writing and increasingly thereafter in polite speech. Most English speakers used heavy dialects, which did not readily succumb even to universal education in Standard English after the legislation of 1870 in England and 1872 in Scotland (Leith, 1983; Wakelin, 1977).

After nearly a century of increasing scholarly attention, the celtic languages were recorded in the Population Censuses of Scotland from 1881 and Wales from 1891. Data on Gaelic were also collected in the Scottish Statistical Accounts of 1791–9 and 1831–45. Most pre-census information is derived from ecclesiastical reports of the language used in church services in areas of Gaelic and Welsh speech, which became common in the eighteenth century (Pryce, 1978a; Withers, 1984). Systematic study of English dialects only began around 1890 (Wakelin, 1977).

Sharp language-divides cannot exist. Neighbours must always be able to talk to each other. Dialects blended imperceptibly, whilst those at opposite ends of England were unintelligible to each other. English merged with the celtic languages over wide zones where there were few monoglot communities. Both languages were pidginized and each was used for different aspects of life. Migrants often used their native language in worship and at home, even when they were proficient in that of their host community (28.1), and taught it to their children (Withers, 1985). Linguistic patterns on the ground are more complex than those which can be depicted on maps.

Standard English and Scots emerged to serve the administrative needs and cultural aspirations of increasingly centralized Tudor and Stuart states, national protestant churches and economies increasingly articulated through mercantile primate cities. English and for a time Scots language aggression had the twin objectives of spiritual enlightenment and political subjugation (Durkacz, 1983; Aikin and McArthur, 1979). Many voluntarily learned English in order to benefit from commercial or distant opportunities and to lose the stigma attached to celtic speech. This erosive force accelerated rapidly through the eighteenth and nineteenth centuries, as did the destruction of the isolated, introverted and self-sufficient ways of life that sustained celtic cultures and languages. These pressures more than compensated for the increasing use of celtic religious texts and the teaching of celtic literacy to provide access to them.

Norn and Cornish succumbed long before the industrial revolution. Except for the towns and gentry Manx was used throughout the Isle of Man in the early eighteenth century. It was used in all churches in at least three out of four services in the early nineteenth century. By 1840, teaching had stopped and Manx was used in only one church service out of four. In the early 1870s, its use was limited to the remote pockets shown on Map 28.5 (Jenner, 1875).

Scottish Gaelic was heavily persecuted for political and missionary reasons. The apparent stability of the Gaidhealtachd through the eighteenth and nineteenth

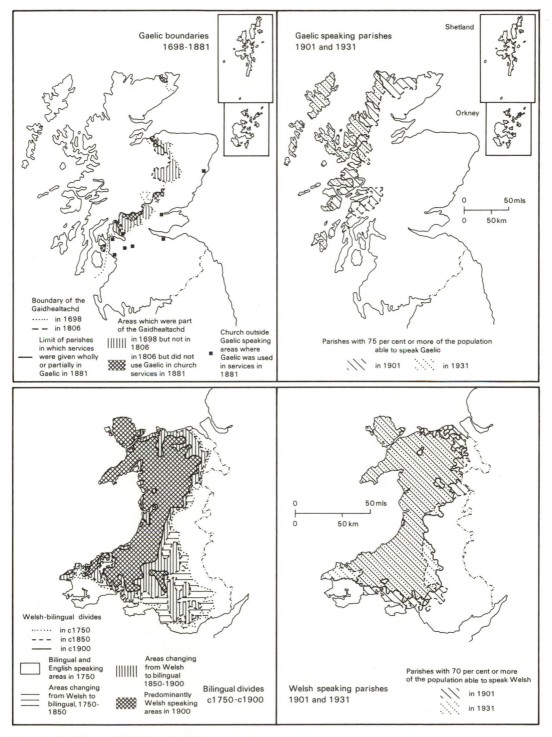

Gaelic boundaries 1698-1881

Boundary of the Gaidhealtachd
...... in 1698
– – in 1806
Limit of parishes in which services were given wholly or partially in Gaelic in 1881 ——

Areas which were part of the Gaidhealtachd
||||| in 1698 but not in 1806
in 1806 but did not use Gaelic in church services in 1881

Church outside Gaelic speaking areas where Gaelic was used in services in 1881 ■

Gaelic speaking parishes 1901 and 1931

Shetland

Orkney

Parishes with 75 per cent or more of the population able to speak Gaelic
in 1901
in 1931

Welsh-bilingual divides

...... in c1750
– – in c1850
—— in c1900

Bilingual and English speaking areas in 1750

Areas changing from Welsh to bilingual, 1750-1850

Areas changing from Welsh to bilingual 1850-1900

Predominantly Welsh speaking areas in 1900

Bilingual divides c1750-c1900

Welsh speaking parishes 1901 and 1931

Parishes with 70 per cent or more of the population able to speak Welsh
in 1901
in 1931

28.1–4 Gaelic and Welsh speaking, *c.* 1700–1931

centuries shown in Map 28.1 is misleading. Although the linguistic and cultural validity of the Highland Line continued to be recognized, there was massive erosion behind it, especially after clan structures and the power of the chiefs were destroyed after 1745. The resulting massive social and economic changes eradicated the economic basis of celtic culture, as did the population clearances and the influx of lowlanders who were more useful to landowners in the new economy (Withers, 1984). Outmigration carried Gaelic to many parts of the Lowlands, and to industrial settlements in parts of the Highlands where its usage was defunct, like the Black Isle (28.1). Glasgow became a major centre for the proselytization of Gaelic through publishing and the activities of the Highland Society, founded in 1727, and the Gaelic Schools Society, founded in 1818 (Withers, 1985). Despite this, the use of Gaelic as a majority language shrank considerably by 1901 and again by 1931 (28.2.).

The persecution of Welsh for political reasons was less intense. Its early use in worship precluded the need for anglicizing missions. The efflorescence of Welsh Calvinistic Methodism in the late eighteenth and nineteenth centuries carried the language into anglicized areas (Pryce, 1975). Because Welsh, unlike Gaelic, was the focus of burgeoning *national* identity, the support from expatriates, especially the London Welsh from the mid-eighteenth century, was much stronger. Most important, Welsh continued to be the language of economic life in Wales. The small farm pastoral economy survived in the north and west and Welsh was used in most of the mining and metal-working regions. Many immigrants into those areas were Welsh speaking. Even the explosive growth of mining in Glamorgan after 1850 did not immediately destroy the language because many of the predominantly English migrants had to learn Welsh in order to work in the pits (Southall, 1892). But their wives and children did not, and after 1870 only English was used in the schools. Hence the erosion shown in Map 28.3, but the population of Glamorgan was still 44 per cent Welsh speaking and the county contained 37 per cent of all Welsh speakers in 1901 (Pryce, 1978b). There was little subsequent shrinkage of Welsh Wales before 1931 (28.4.).

Vigorous efforts to spread Standard English in the eighteenth century and after were countered by exhortations to retain dialect speech (Wakelin, 1977) and by superb literary accomplishments in dialect. Written political debates were often couched in dialect in broadsheets and newspapers, with 'outside' views, as often radical as not, stigmatized by expressing them in Standard English (Langton, 1984). Publication in dialect increased rapidly through the nineteenth century, especially after 1850 (28.6). This did not occur in all the industrializing areas of England, as the Midlands show. There were, too, great variations of language and practice even across the north. The translations of the *Song of Solomon* shown on Map 28.6 recognized different dialects within Lancashire and Yorkshire, and Edwin Waugh, the foremost 'Lancashire' dialect poet, was more at home in textile districts outside Lancashire and in towns to which textile workers had migrated in the cotton famine of the 1860s than he was in southwest Lancashire (28.7). Dialect poems and dialogues were meant to be read aloud, to give vivid expression to the extremes of coarse jollity, painful sadness and intense personal affection to which dialect pronunciation, vocabulary and grammar are particularly suited. The self-conscious use of broad dialect was not simply a means of working-class identification. It expressed intense localism and a preoccupation with personal feelings and intimate relationships, a set of values far removed from the dutiful nationalist conformity, the 'stiff upper lip' and objective empiricism of metropolitan English culture.

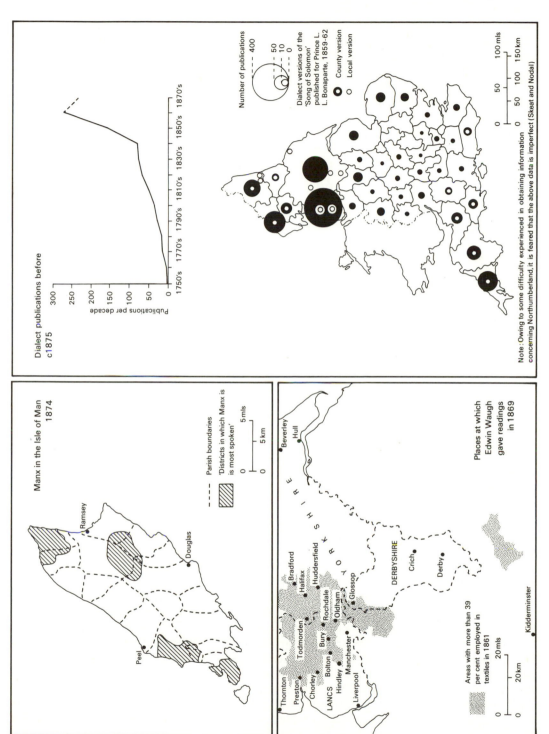

Number of publications

— 400

— 50
— 10
— 0

Dialect versions of the
'Song of Solomon'
published for Prince L.
L. Bonaparte, 1859-62

● County version
○ Local version

Note: Owing to some difficulty experienced in obtaining information
concerning Northumberland, it is feared that the above data is imperfect (Skeat and Nodal)

Dialect publications before
c1875

Publications per decade

300
250
200
150
100
50
0

1750's 1770's 1790's 1810's 1830's 1850's 1870's

Manx in the Isle of Man
1874

Ramsey

Douglas

Peel

- - - - Parish boundaries

▨ 'Districts in which Manx is
   is most spoken'

0   5km
0   5mls

Beverley
Hull

Y O R K S H I R E

Bradford
Halifax
Huddersfield
Thornton
Preston
Todmorden
Rochdale
Bury
Oldham
Glossop
Chorley
Bolton
Manchester
Hindley
Liverpool

DERBYSHIRE
Crich
Derby

LANCS

Kidderminster

Places at which
Edwin Waugh
gave readings
in 1869

▨ Areas with more than 39
   per cent employed in
   textiles in 1861

0   20km
0   20mls

28.5-7 Manx speaking, 1874; English dialects in the nineteenth century

29 Education

W.E. Marsden

The important, though neglected, spatial aspects of the development of schooling are here considered in relation to England and Wales, whose educational system was formally separate from that of Scotland. These aspects were diffusion; regional variation (Marsden, 1982a; Stephens, 1973); scale, whether national, regional/metropolitan or local/community (Marsden, 1977a); and social differentiation (Marsden, 1983).

In the process of educational development, different criteria can be used as indices of achievement. As far as formal education is concerned, provision of accommodation provides a baseline. The next index is the actual enrolment of children, above which are levels of attendance, best measured in terms of average attendance over a year. A linked criterion is length of stay at school, an important influence on the culminating measure, standards achieved. The most basic and, it must be said, not altogether convincing, criterion of achievement was literacy (Stephens, 1977) as measured by ability to sign the marriage register, an index of declining validity as the century and mass provision progressed. More appropriate measures for the later decades of the century in the elementary sector were the standards reached under the Codes laid down by the Education Department. Whatever the criterion used, regional variations abounded and persisted.

The Brougham-inspired surveys of the education of the 'lower orders' in the 1810–20 decade are an important source on comparative provision over England and Wales (*BPP* 1816, IV). The foremost parliamentary educational reformer of his time, Brougham sought to introduce mass elementary schooling on Scottish parochial lines. The 1818 surveys brought to light glaring variations in uptake (29.1) (*BPP* 1818, III, IV). The relatively advanced state of the northern counties was striking. Contemporaries ascribed this to the healthy influence of Scotland, an example of contagious diffusion. At the other extreme were the Welsh counties and, the disparity which most exercised Brougham, the two most densely populated counties, Lancashire and Middlesex. '

The educational societies, the National (Anglican) and the British (non-sectarian), were established to disseminate the 'new system' of mass provision, that is the monitorial system. Receiving Treasury grants from 1833, and joined later by the Catholic Poor Schools Society, these voluntary agencies sought to plug gaps, most particularly in the large towns and cities. But it was in these that the demands were greatest, the resources least adequate, and the threshold of success insufficient to prevent state intervention.

There had been some improvement by 1851, on the basis of children attending day schools on census day (29.2). The Anglican heartlands of southeast England had made conscious advances. But Wales, Monmouth and Hereford remained backward, as did the Black Country, Lancashire and London. A comparable situation existed in Scotland, though its reputation as being in advance of England and Wales tends to be supported by these figures. It had, however, its lagging regions: heavily urbanized Clydeside and the remote areas of the northwest, and Orkney and Shetland. There appears to have been some slippage, relatively speaking, in northern England, perhaps associated with industrialization in the northeast.

Intra-regional variations are also highlighted by the 1851 educational census. It was widely agreed that the demographic imperatives of the large cities posed the greatest problems for the voluntary providers. Within these too, resources were in the places

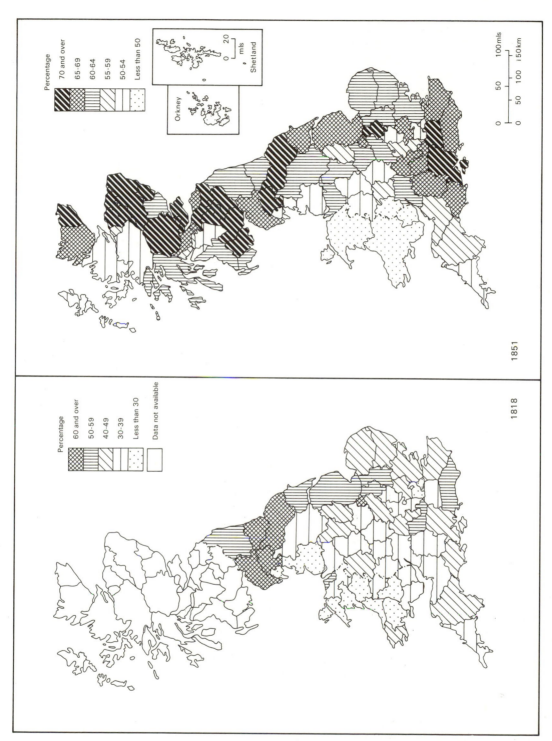

29.1–2 Children in schools, 1818 and 1851

where they were least needed. Maps 29.3–6 show well the advantaged and disadvantaged parts of the metropolitan area, with higher attendances in the suburbs of the northwest, southwest and southeast, and lower in the East End and the slums surrounding the city centre. This index of course conveys no hint of the differential quality of the provision between these areas.

The relative failure of the voluntary societies to resolve the problem of mass urban provision was essentially the justification for the great Education Act of 1870, which allowed the use of the rates for educational provision. School boards were established to 'fill the gaps'. Under them attendance was improved, and by 1880 the compulsory principle was enshrined in the legislation.

The voluntary agencies presumed the role of the school boards would be to concentrate on providing for 'ragged' children, leaving to them the more congenial task of educating the respectable. This was not to be the school boards' definition of the situation. The main problems faced immediately by boards such as London's were however those of poverty, and more especially the territorial concentration of poverty (Marsden, 1977a) as in Southwark, the East End and the northern fringes of the City and Westminster. The accommodation problems faced in 1871 are shown on Map 29.4. Almost all districts were short of infant places, but the greatest overall deficiencies were in the poor areas. The City and Westminster, with long-standing endowments and, in the case of the City, declining rolls, were relatively well placed. As Map 29.5 indicates, the needy areas were those in which the school board provided most. It must of course be stressed that there were significant pockets of poverty in the generally more prosperous districts, such as Greenwich and West Lambeth. Board school provision was least in the City, Westminster and Marylebone.

In addition to this work for the poorer groups, some urban boards were bent on providing for the respectable. London's presence was felt in most social areas (29.5). Although church schools undoubtedly retained residual status, even the more comfortable parents could not but be impressed by the new board schools, some of the finest elementary buildings ever constructed. The resources of the rates also allowed better equipment, better teachers and, most important, the erection of a scholarship ladder (Marsden, 1982b). Map 29.6 illustrates the variations of achievement in this area at the turn of the century by measuring the proportion of candidates who had reached an elementary standard high enough to enter for scholarships, again showing the northwest, southwest and southeast districts as advantaged.

Ecological relationships between schools and their communities can best be displayed at the micro-scale of study, in this case from two examples of slums and suburbs in west London (Notting Hill) and north Merseyside (Bootle). Thus in Notting Dale (Map 29.7), identified in Booth's surveys (vol. 2, 1891) as one of the blackest holes of poverty in London, only one-third of the population was 'in comfort', in contrast to nine-tenths in adjoining Holland Park and the Ladbroke Grove area. In the latter, children did not of course use the elementary schools. But even away from these affluent districts, there was still seen to be a need to segregate schools, so that the children of respectable working-class and lower-middle-class parents could be quarantined from those raised in squalor and vice. The mechanism was the graded school fee. At St Clement's Board School, in the middle of the Notting Dale slum, the fee was the minimum 1d. per week, and many of its children were drawn from streets immediately around it. By contrast, the nearby Saunders Grove (later Road) Board School, conveniently placed also for this

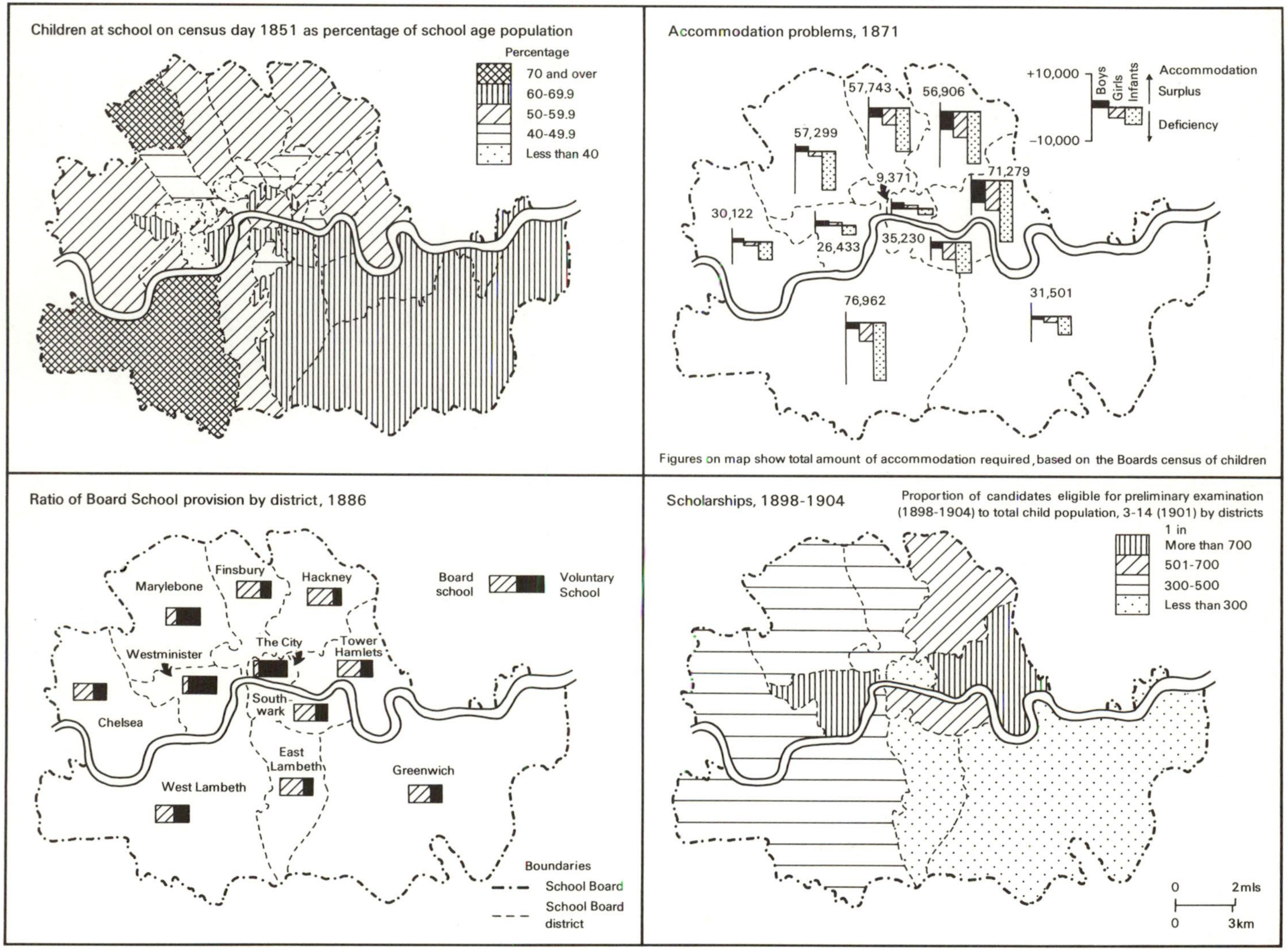

Accommodation problems, 1871

Accommodation
Surplus
Deficiency

Boys
Girls
Infants

+10,000

−10,000

57,743
56,906
71,279
9,371
35,230
57,299
26,433
76,962
30,122
31,501

Figures on map show total amount of accommodation required, based on the Boards census of children

Scholarships, 1898–1904

Proportion of candidates eligible for preliminary examination (1898–1904) to total child population, 3–14 (1901) by districts

1 in
More than 700
501–700
300–500
Less than 300

0   2 mls
0   3 km

Children at school on census day 1851 as percentage of school age population

Percentage
70 and over
60–69.9
50–59.9
40–49.9
Less than 40

Ratio of Board School provision by district, 1886

Board
school

Voluntary
School

Marylebone
Finsbury
Hackney
The City
Tower Hamlets
Westminster
South wark
East Lambeth
Greenwich
Chelsea
West Lambeth

Boundaries
School Board
School Board
district

29.3–6   London schooling, 1851–1904

area, omitted to take any children from it. In the middle-class fringe area of Oxford Gardens, parents petitioned the London School Board to levy the high fee of 6*d*., to keep the school select.

Bootle also illustrates well the impact of social segregation, both between well-to-do middle-class groups and the working classes, and within the working-class group. Indeed, the respectable working class had more in common with the lower middle class than they had with the 'disreputable poor'. In Bootle, by 1895, there was roughly a middle-class enclave; two slums adjoining the docks, an old-established one in the southwest and a more recent one in the northwest; and otherwise reasonably respectable working-class terraced streets (Map 29.8). In the dockland areas lay the low-prestige schools, in particular St John's (Anglican), St Alexander's (Catholic), both established just before the school board was formed, and Salisbury Road (a board school). St James's (Catholic) also took large numbers of children of casual dock labourers, but set up in a separate building a 'select' school for the children of shopkeepers and the like. St Mary's (Anglican), Bootle's first National school, enjoyed a good reputation until dockland development caught up with it. It retained some residual prestige, but many of its more respectable children were moved to the middle-class fringe Christ Church school, which achieved an enviable reputation in the town (Marsden, 1977c). Two board schools, Bedford Road and Hawthorne Road, also enjoyed a good reputation through their location in respectable working-class areas. Children were regularly transferred from school to school, often for capricious reasons. At one time almost a triangular trade grew up: Christ Church offloading troublesome children on to Salisbury Road when it opened, while another group of children was shifted from Salisbury Road to St Mary's; then others, as already noted, moved from St Mary's to Christ Church. Perhaps the most telling statement of the macro-divide is illustrated by the residential segregation of the predominantly middle-class members of the Bootle School Board (Marsden, 1978) who sat on the tribunals which dealt with the parents of children summoned for non-attendance, drawn mostly from the southwest slum area. Otherwise they were not to be seen crossing the physical barriers between the social zones provided by the Liverpool–Southport railway and the Liverpool and Leeds canal. But it must not be forgotten that within the working-class/lower-middle sector another more finely-tuned set of gradations in schooling faithfully reflected the territorial segregation of the social groups brought in its train by urbanization.

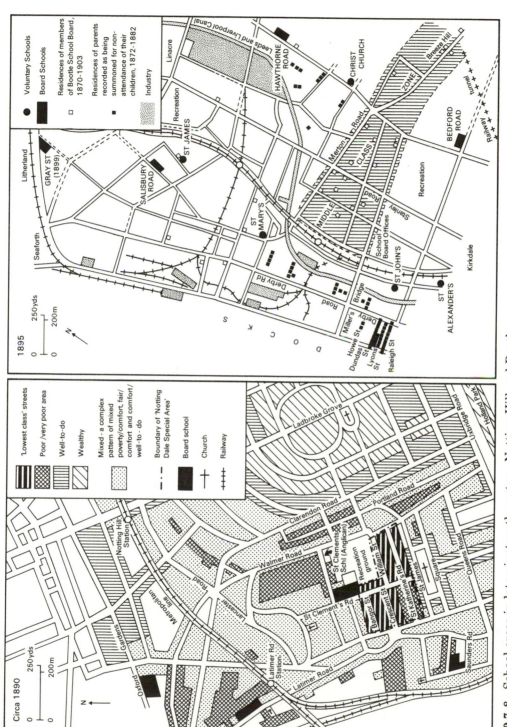

29.7–8 School areas in late nineteenth-century Notting Hill and Bootle

30 **Religion**
Hugh McLeod

The first and only British census of church attendance took place on Sunday, 30 March 1851. Interpreting the figures needs care. Sunday school and congregational attendances were not distinguished. Many attended two or even three services in the day, leading to considerable double counting. The non-returns were unevenly spread across the country, being especially high in Scotland. Thus Scotland is treated separately in Map 30.1. Within these limits the returns are substantially accurate.

In all three countries there were enormous variations in the level of church-going. In England, the ratio of total reported attendances to population varied from 16 per cent in the Longtown Registration District (Cumberland) to 116 per cent in St Ives (Hunts.). In Wales it was from 36 per cent in Knighton (Radnor) to 121 per cent in Aberystwyth (Cardigan). In Scotland, it was 14 per cent in rural areas of Selkirkshire, but 132 per cent in Dingwall (Ross). In all three countries the lowest attendances were in rural border regions. Although attendances were substantially higher in Wales than in England, the basic patterns of attendance were similar, and in two important respects different from the Scottish. First, regional differences in the level of attendance were more sharply defined in the former countries, and were reflected both in town and country. Second, attendances were higher on average in rural than in urban areas, whereas in Scotland the reverse was true. English and Welsh districts with very high levels of attendance were predominantly agricultural, though they included some areas of industrial villages and small towns; very high attendances were rare in towns of over 12,000 people (McLeod, 1973).

One of the most significant divisions in nineteenth-century Britain was that between 'church' and 'chapel', the former being adherents of the Established Church of their country, and the latter being Protestant Dissenters (Gilbert, 1976) (30.2). The best guide to their relative strength is the 1851 census of church attendance, though the census may over-represent 'chapel' strength because of the greater regularity with which Dissenters attended services. As far as attendances went, 'church' was slightly ahead of 'chapel' in England, while 'chapel' was far ahead in Wales (I. G. Jones, 1981) and comfortably ahead in Scotland. In Scotland the three leading denominations, one established and two dissenting, were of roughly equal strength. Map 30.3 therefore simply shows the largest denomination in each burgh and in the rural areas of each county. But in most parts of England, and even in some parts of Wales, the Anglicans were by far the largest denomination, and the major question was their strength in relation to the combined forces of Dissent. The denominational pattern was also more variegated than in Scotland. For England and Wales Maps 30.2–3 show two things: the relative strength of 'church' and 'chapel' in each district, and the largest non-established denomination. Support for the Established Churches was highly regionalized (Gay, 1971). The Church of Scotland was very weak in the Highlands but relatively strong in the east. Anglican strength was heavily concentrated south of the Trent, east of the Severn and east of the Tamar. Anglican predominance in much of the west Midlands and northwest reflected the paucity of Dissenters in regions where church-going of all kinds was low. Dissenting strength was more variable, and also less locally consistent and less regionalized. They were ahead in most towns; in rural areas chapel attendance ranges from very low to very high. They could reach people who would never go to the church; but 'church' could always rely on a minimum of support, where 'chapel' sometimes made no impact at all.

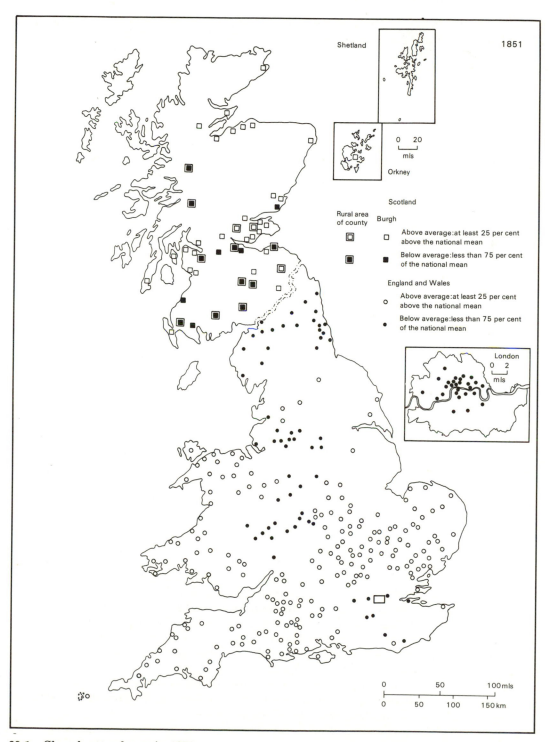

30.1 Church attendance in 1851

The characteristic religious movement of the English industrial revolution was Methodism. Beginning as a revival movement in the Church of England, Methodism gradually established its own completely separate organization and identity (Ward, 1972). By the early nineteenth century it was the predominant form of religion in large parts of northern England. The Methodist societies were organized in a system of circuits. Maps 30.4–6 trace the growth of Methodism by showing the location of circuits. Methodism first became established in the areas where the Anglican Church was most conspicuously failing to respond to the rapid growth of population – in the larger towns, and in the textile and mining districts. The basic features of nineteenth-century Methodist geography were already apparent when the first statistics were published in 1767. Of the 23,000 British Methodists in the Connexion established by John Wesley, 48 per cent lived in Yorkshire, Lancashire or Cornwall. Lincolnshire was established as the most strongly Methodist of the agricultural counties. The pattern had changed little by 1791, when Wesley died (30.4). In the next fifty years the followers of Wesley split into several separate denominations, and the number of British Methodists increased from 57,000 to 459,000. Maps 30.5–6 show the growth of the main body, the Wesleyan Methodists. Map 30.7 shows the location in 1841 of the smaller Methodist denominations. Between 1791 and 1821 Methodism became a nation-wide movement. Progress remained steady in the older strongholds, but there was spectacular growth in secondary areas like the north Midlands, and in former 'Methodist deserts' like Wales and southeastern England. Only in Scotland did support remain highly localized. The period 1821–41 was more one of consolidation, when the ratio between Methodist membership in different regions stabilized. In 1841, 36 per cent of Methodists still lived in the three major counties. The smaller denominations were particularly concentrated, except for the Primitives, who broke new ground by winning support in strongly Anglican rural areas.

The overall level and denominational distribution of church attendance varied considerably within nineteenth-century cities (30.8–11). This was already apparent in 1851 when separate figures were published for each of the thirty-six registration districts in London. Increasingly clearly defined residential segregation made similar analyses possible in other cities when a series of counts of church attendance were organized by local newspapers between 1881 and 1912. The best of these was the *Daily News* census of London, conducted over several Sundays during 1902 and 1903 (McLeod, 1974). The most important differentiating factor was class, here indicated by the proportion of population living in households of six or more rooms (30.8). The overall level of church-going tended to be considerably higher in districts with a large middle- or upper-class element than in those that were heavily working-class. The relationship between class and religious practice was strongest in the case of Anglicanism; Nonconformists, on the other hand, while being very weak at either end of the social scale, were distributed relatively evenly within a wide social band extending from the upper working class to the upper middle class. The pattern of Nonconformist strength was also complicated by a regional factor: the chapels were generally strong in the northern and northeastern suburbs of London.

Nearly all cities had their largest congregations in middle- and upper-class areas, the exception being Liverpool, where, in the early twentieth century, attendances were as high in Catholic working-class areas as in the suburbs.

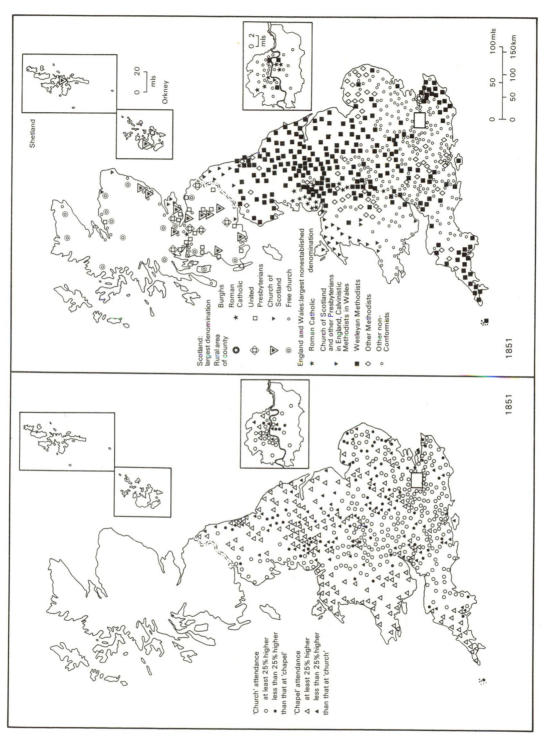

Scotland: largest denomination

	Rural area of county	Burghs
Roman Catholic		
United Presbyterians		
Church of Scotland		
Free church		

England and Wales: largest nonestablished denomination

Roman Catholic

Church of Scotland and other Presbyterians in England, Calvinistic Methodists in Wales

Wesleyan Methodists

Other Methodists

Other non-Conformists

1851

'Church' attendance

○ at least 25% higher

● less than 25% higher than that at 'chapel'

'Chapel' attendance

△ at least 25% higher

▲ less than 25% higher than that at 'church'

1851

30.2–3 Denominational strength, 1851

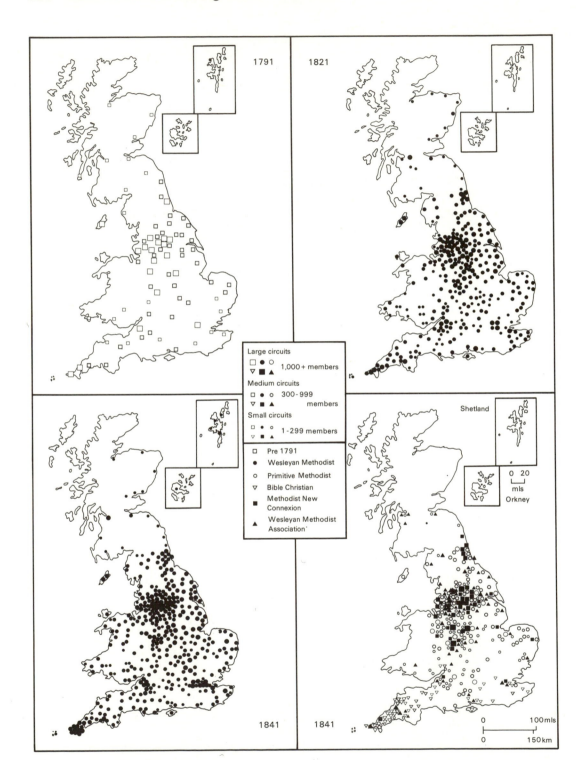

30.4–7   Methodism, 1791–1841

1911

Percentage living in households occupying 6 or more rooms

45.6 - 60.0
32.1 - 43.7
22.9 - 31.6
8.0 - 21.3

1902 - 3

Attendance by adults at churches of all denominations as a percentage

30.3 - 47.4
23.2 - 27.0
18.3 - 22.8
11.8 - 17.7

1902 - 3

Attendances by adults at services of the Church of England as a percentage

15.4 - 28.5
10.5 - 14.9
7.4 - 9.8
4.2 - 6.8

1902 - 3

Attendances by adults at Non-conformist chapels as a percentage

13.0 - 22.6
9.3 - 12.1
7.8 - 8.7
4.2 - 6.8

0   5 mls
0   5 km

30.8–11 Church attendance and status in London, early twentieth century

31 The electoral system

Michael Hurst

The geographically dispersed aspect of the political system, namely the election of members to the House of Commons, was only one aspect of a larger structure. Laws were the work of 'The King (or Queen) in parliament'. The Crown still retained great political clout when the industrial revolution began. It could certainly destroy cabinets, although frequently unable to maintain them. Only in 1835 was it obliged to accept wholesale one devised by a parliamentary leadership in control of the Lords and Commons. Great influence (actual or potential) was commanded by the House of Lords *vis-à-vis* the House of Commons throughout the late eighteenth, the nineteenth and the early twentieth centuries. Since the 1670s it had lost the power to veto money bills, but not until 1911, with the passing of the Parliament Act, was its right to reject all other prospective measures subjected to severe restriction. Contemporaneously, for long into the nineteenth century and occasionally into the twentieth, peers were able to exercise enormous influence over the conduct and outcome of elections to the Commons. The great reality of power from above must never be left out of account.

These parliaments were elected by a multi-national country. And right through our period, first England, Wales and Scotland, and then (ultimately) Ireland too, all had singular features to their electoral methods. The most important related to the constituency systems and the franchises employed. England was the great originator of what became the bicameral legislature of the United Kingdom. Wales was incorporated into it in 1536, Scotland in 1707 and Ireland in 1800. While the constituency and franchise patterns of England were comparatively confused and collectively irrational long before 1832 and the first 'Great Reform' of Parliament, those of the peripheral states were more systematic from the outset. A simple situation in the case of Wales, and the need to shrink existing institutions in those of Scotland and Ireland, worked powerfully in that direction. All four countries had two types of Commons election units – counties and boroughs. In the unreformed days English and Irish counties had (Yorkshire excepted) two members; whereas one only was the rule in the Welsh and Scottish equivalents. At the same time England, Wales and Ireland initially enjoyed almost identical franchise laws, variously interpreted in the light of discrete conditions. But in 1829 a much restricted voting law came in for Ireland as a concomitant of the completion of 'Catholic Emancipation' and destroyed what was almost certainly the widest franchise then operative in Europe. Scotland was odd man out, with a franchise narrower than that of Ireland in its more constricted phase. While all but five of the 203 English boroughs, apart from London with 4, had 2 members each, all Welsh and all Irish, save Dublin and Cork, sported but 1. Again Scotland stood on its own. Some small towns in Wales were bunched together to constitute a borough seat. Paucity of large centres had dictated it. Edinburgh alone among Scottish urban areas returned a member of its own. The rest were grouped, with the members returned by indirect election – a remarkably singular feature (31.1). Borough franchises were of many varieties. Many boroughs approached close to universal suffrage, though the generous importation of outsiders in the guise of freemen sometimes undermined the freedom of choice that this should have brought. Borough councils of numerous places returned the members – the invariable initial basis of election in Scotland and another proof of its distinct position. A few towns had electorates built up from the owners of particular sites. Others, the 'rotten boroughs' had few or even no electors at all.

The leading concept behind parliamentary representation before 1832 was that of the

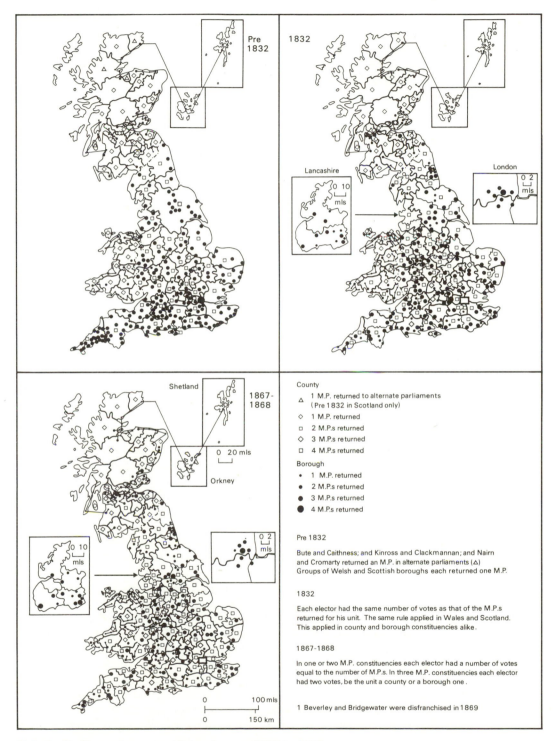

31.1–3   Parliamentary constituencies, pre-1832 to 1868

Within the figure:

Pre 1832

1832

1867-1868

Lancashire

London

Shetland

Orkney

County

△ 1 M.P. returned to alternate parliaments
   (Pre 1832 in Scotland only)

◇ 1 M.P. returned

□ 2 M.P.s returned

◇ 3 M.P.s returned

□ 4 M.P.s returned

Borough

• 1 M.P. returned

• 2 M.P.s returned

● 3 M.P.s returned

● 4 M.P.s returned

Pre 1832

Bute and Caithness; and Kinross and Clackmannan; and Nairn
and Cromarty returned an M.P. in alternate parliaments (△)
Groups of Welsh and Scottish boroughs each returned one M.P.

1832

Each elector had the same number of votes as that of the M.P.s
returned for his unit. The same rule applied in Wales and Scotland.
This applied in county and borough constituencies alike.

1867-1868

In one or two M.P. constituencies each elector had a number of votes
equal to the number of M.P.s. In three M.P. constituencies each elector
had two votes, be the unit a county or a borough one.

1 Beverley and Bridgewater were disfranchised in 1869

economic interest group. Substantial population generally but not necessarily went along with the idea that an area was a meaningful source of potential taxation. Such taxation, once deemed important, was not looked at in a precise way. Only when yield reached a massive level was any variation of representation entertained. While all English counties and most English boroughs were given two members each, only London and Yorkshire ever got more. The English county franchise of the 40-shilling freeholders meant an electorate uniform in types of person but not in density of distribution. Widely differing borough systems necessarily resulted in hotchpotch conditions. Not, however, until the industrial revolution was well developed did the actual distribution of borough (unlike county) seats become seriously inadequate. For while certain places had become defunct or virtually so, the large majority flourished and very few substantial centres lacked separate representation. Even then more already existing boroughs became crucial industrial centres than places which first received seats in 1832. Before 1800 borough seats made up 78 per cent of the Commons; 1800–32, 71 per cent; 1832–68, 62 per cent; 1868–85, 57 per cent; and 1885–1918, 44 per cent.

Parliamentary reform had the dual effect of knocking out separate representation for units deemed unfit at the various stages of change, and of introducing it for ones thought more deserving (31.1–3) (Cannon, 1973). With each shift the concept of economic interest groups becomes progressively supplanted by one of population numbers, turning gradually into one of electorate size. The English counties gained and small boroughs increasingly disappeared. The constituency system was thus revolutionized. Single-member constituencies became the virtual overall rule by 1885 (31.8), at the point where the near-democracy of the borough seats brought in with the 1867 Act became matched in those of the counties. As industrialization spread, so numerous one-member county constituencies meant more effective representation for modern society than numerous twin or single-member small boroughs. With by far the largest number of members, England (pre-1800, 88 per cent; 1800–32, 74 per cent; 1832–68, 72 per cent; 1868–85, 70 per cent; and 1885–1918, 70 per cent) took the key role. Wales's borough proportion fell from 50 to 37 per cent. Scotland alone saw the status quo virtually maintained. It, in fact, had been the big overall gainer of seats at every reform. England had 489 seats pre-1832; 471 in 1832; and 465 in 1868 and 1885. Wales had 24, pre-1832; 29 in 1832; and 30 in both 1868 and 1885. Ireland's movement was yet more modest – 100, pre-1832; 105 in 1832; 105 in 1868; and 103 in 1885. But Scotland had shot up from a pre-1832 figure of 45 to 53 in 1832, followed by a jump to 60 in 1868 and to 72 in 1885. This upward trend should rightfully have been balanced by a downward one for the south and west of Ireland.

For Britain, 1832 brought a more varied yet still relatively uniform county franchise; for Ireland merely an amelioration of the cutbacks of 1829, though in 1850 a more radical change brought about a far more consistent United Kingdom pattern. In the boroughs a basically uniform franchise (based on the 10-pound householder) was created for the first time. The electorate increased by approximately 50 per cent. Before the 1867 Act for England and Wales and those of 1868 for Scotland and Ireland, growth in national wealth, political energy and the Irish Act of 1850 brought an increase of a further 60 per cent. Near borough democracy and the adding of smaller leaseholders to county lists doubled the electorate in 1867 and 1868. The advent of near democracy in the counties during 1884 and 1885 hiked the figures up yet another 75 per cent to 80 per cent. Proper voter lists had been inaugurated in 1832. Strigent tax payment and residence qualifications had always to be satisfied before any person could be registered as a voter.

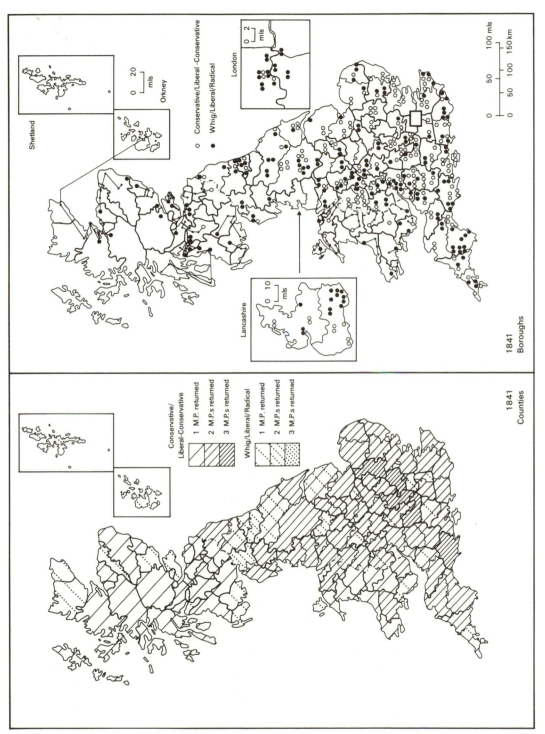

31.4-5 Election results, 1841

In 1913 it was estimated that fully 20 per cent of ostensibly eligible males of 21 and over were thereby excluded from exercising the franchise; something much to the grave disadvantage of the less affluent sections of the people. It should, nevertheless, be understood that even in 1832 the bulk of what would now (1985) be regarded as middle and upper working-class males of appropriate age did stand an excellent chance of becoming voters. On the other hand, the retention of means of having more than one vote through qualifying in different places for different properties certainly profited the better off. Even in the early twentieth century such persons have been estimated to have numbered 250,000 and more in an electorate of just over 7.7 millions.

Corruption was for long a widespread curse of elections (Gash, 1952). Only an Act of 1883 effectively curbed it and its effects. But with the widening of the electorate and the ever-increasing politicization at work, cost and ideology had hindered its operations. Contrary to widespread belief, politics throughout the eighteenth century were deeply politicized at the top (and according to area at various levels below it). Unopposed returns were very numerous even in December 1910, but did not usually signify political indifference or inactivity. Both the Jacobite and Jacobin problems created politics of an unhealthy lopsidedness, which rendered opposition weak for long periods. Yet before the first arose and for long after the second disappeared (not to mention the interim between them) astute calculation of changes was the supreme dictator of decisions to contest or not to contest constituencies. While true that party authority was highly localized before 1832 and strongly so in many places for long thereafter, distinctly national trends can be easily discerned in election results, both as to victory or defeat and to rises or falls in votes won.

Anomalies were legion throughout our period. Never was the property or the population principle applied with near absolute consistency. An especially puzzling feature was that until 1918 every borough constituency was sitting in a county one. Thus men residing in the towns frequently qualified for and voted in county elections by dint of having properties not used for a borough vote. So, while true that Birmingham and Manchester were represented after 1832, many within their boundaries voted after as before for their respective Warwickshire and Lancashire county members. Most crucial of all, do note, that warts and all, every constituency and franchise system employed did lend itself to a full working of a two or more party system. General elections did change governments under all of them. However much the system of pre-1832 became jammed by the anti-Jacobin reaction, calmer times brought fluidity. Lord Grey's Ministry was essentially the child of an unreformed electorate and the 1832 Act the work of an unreformed constitution. Ironically enough, the percentage of the votes given to a particular party in a particular place frequently remained near constant whatever the franchise used. Virtual and actual representation did therefore often coincide.

The election results chosen illustrate the changing impact of the electoral system as well as of the changing economic and social structure. The 1841 general election was the first election to be directly responsible for the replacement of one government by another (31.4–5). It was a conservative victory by 367 seats to 291. This was the culmination of tory recovery from the defeat of 1832. Peel had rebuilt the party around defence of the Established Church. The maps show conservative strength before the split over the Corn Laws. There was a tendency for a larger proportion of boroughs to be conservative in the east and south of England, the areas of strongest Anglican support.

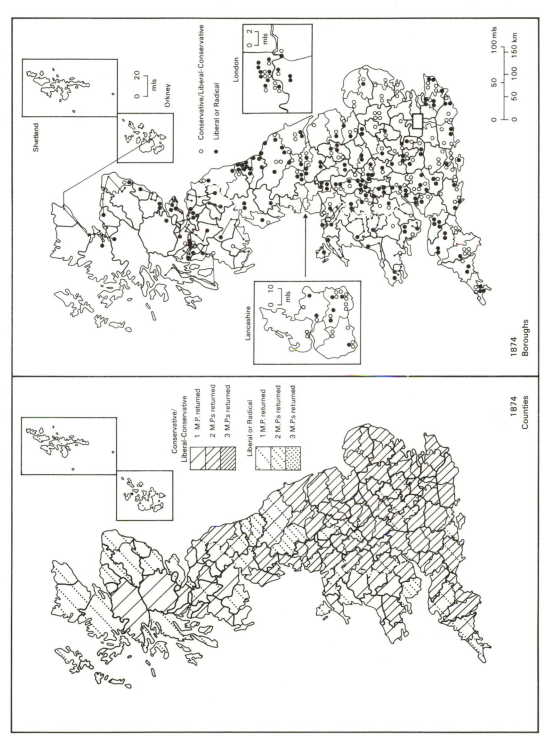

31.6-7   Election results, 1874

Too much should not be read into this for the major influence on the result was often the political views of the dominant economic interest in the constituency, usually the landowners (Hanham, 1959). The most striking feature was the rush back to the conservatives in the English counties. Here there were only 20 liberal seats to 124 tories.

The 1874 election was the first under the secret ballot and the second under the extended franchise of 1868 (31.6–7). The conservatives won 350 against 242 liberals and 60 Irish Home Rulers in an election in which Gladstone claimed he was drowned in a torrent of gin and beer. Certainly the unpopular Licensing Act with its restriction of drinking hours was one cause of his defeat. The maps show the overwhelming predominance of tories in the counties still. In England the line from the Humber to the Exe was now more than ever a division between tory and liberal boroughs. The trend which linked Wales and Scotland to liberal strength was beginning to assert itself.

The 1886 election was fought on the issue of Home Rule for Ireland (31.9–10). It was a major defeat for Gladstone and marked the liberal split. The liberals retained their hold on the Welsh-speaking counties, and many of the urban seats in the north and Scotland. There were several distinctive features. Liverpool and Manchester, like much of south Lancashire, were conservative, with the redoubtable T. P. O'Connor, the sole Irish Nationalist in Britain. Joe Chamberlain had led Birmingham wholesale into the unionist fold. London, generally, showed a strong trend to be conservative and unionist. The divison of the larger northern cities into single-member constituencies was delivering a steady stream of seats to the conservatives.

The 1906 election was a devastating liberal victory, 377 to 157 conservative and unionist and 53 labour (31.11–12). Fear that high food prices would result from conservative belief in protective tariffs swept the liberals to victory. The unionists were swept from Wales and nearly so from Scotland. Labour members began to appear in the coalfields and some working-class seats in large towns, notably London. The conservative survivors were as much in the boroughs as in the counties. Birmingham stood firm and more against the tide; Liverpool, Sheffield and Glasgow (in that order) also stuck remarkably to the unionist cause.

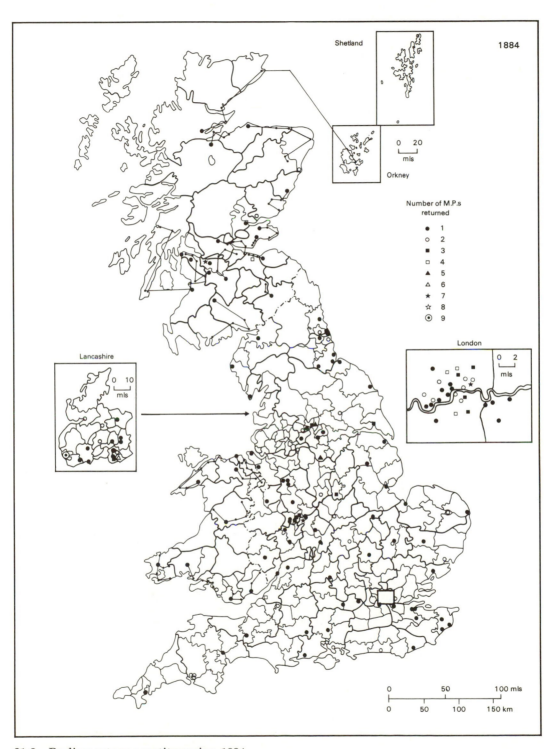

31.8 Parliamentary constituencies, 1884

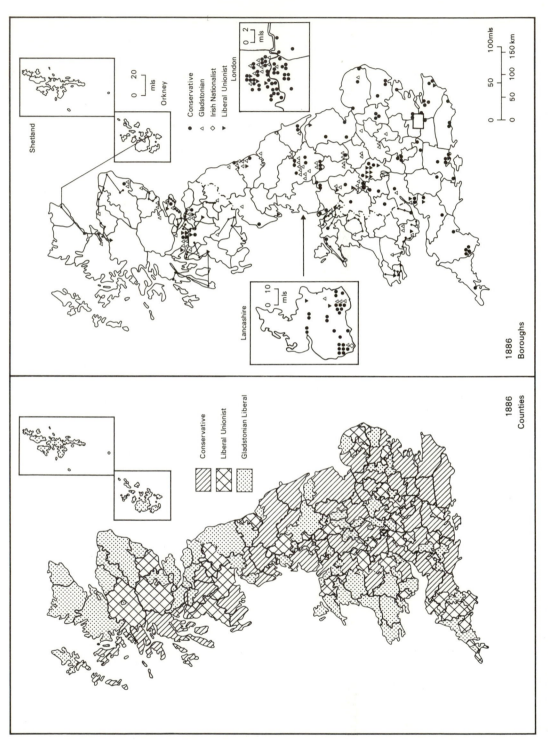

31.9–10   Election results, 1886

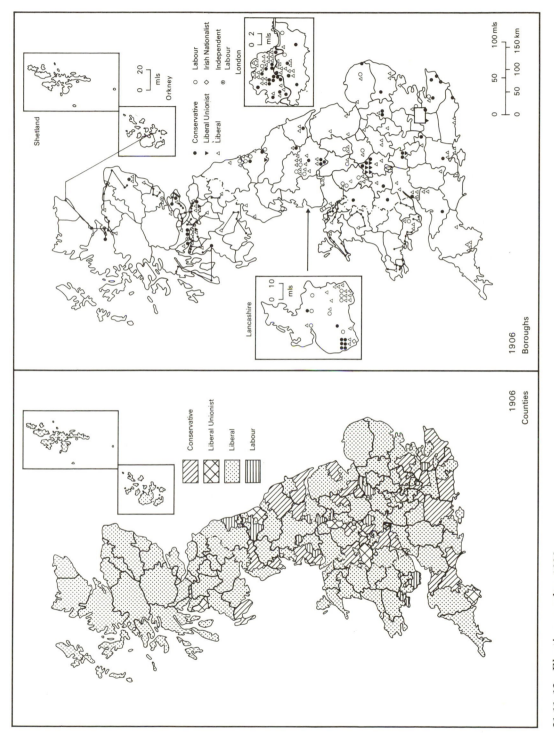

31.11–12   Election results, 1906

Sources of maps

1 The physical environment

1.1 Carter *et al.*, (1974), 33.
1.2 Doornkamp and Gregory (1980), 26.
1.3 Eastwood (1964), 45 and 50.
1.4 Eastwood (1964), 48 and 50.
1.5 Watson and Sissons (eds) (1964), 92.
1.6 Coppock (1971), 35.
1.7 Manley (1974), 393–7 and Wigley *et al.* (1984), 14–15.
1.8 Bilham (1938), 156. The earliest available mapped climatic data have been used in
1.8–14 The years to which statistics relate are given if they are provided in the source.
1.9 Bilham (1938), 157.
1.10 Coppock (1971), 35.
1.11 Coppock (1971), 35.
1.12 Bilham (1938), 251.
1.13 Bilham (1938), 84.
1.14 Coppock (1971), 35.
1.15 Ward (1981), 19.

2 Population

The data for the maps in this section were largely derived from the abstracts of the Census printed in British Parliamentary Papers. For an excellent guide to these see Office of Population Censuses and Surveys, *Guide to Census Reports, Great Britain, 1801–1966*, London, HMSO, 1977. See also Lawton (1978b).

3 Regional structure and change

The data were drawn from the British Census. For the problems and definitions involved in the preparation of the maps see Lee (1979).

4 Agriculture

4.1–20, 4.29–36, 4.39–48 1801: County acreages for individual crops are in Turner, M. E. (1981) 'Arable in England and Wales: estimates from the 1801 Crop Return', *Journal of Historical Geography*, 7, 291–302;

only counties with more than 10 per cent of their area covered by the return have been used. County acreages are from the 1831 Census. Yield information is given in Turner, M. E. (1982) 'Agricultural productivity in England in the eighteenth century: evidence from crop yields', *Economic History Review*, 2nd series, 35, 489–510. A complete transcript of the 1801 Crop Return is in Turner, M. E. (ed.) 'Home Office Acreage Returns (HO 67) List and Analysis', *List and Index Society*, 189 (1982), 190, 195 (1983) and 196 (1983).

1840s/1850s: Crop acreages and yields for England and Wales from Kain, R. J.P. and Prince, H. C. (1984) *The Tithe Survey of the Mid-Nineteenth Century*, Cambridge, 226–8. Only counties with more than 10 per cent of their tithe districts with information have been used. County acreages are from the 1851 Census. For Scotland, data refer to 1854 and are in Maxwell, J. H. (1855) 'Report of the Highland and Agricultural Society of Scotland to the Board of Trade on the agricultural statistics of Scotland for the year 1854', *Highland and Agricultural Society of Scotland, Prize Essays and Transactions*, 3rd series, 6: 483–512.

1871: Acreages and livestock from agricultural returns for Great Britain for 1871, c. 460, BPP 1871, LXIX. Yields are averages for the period 1863–1882 as given by Craigie, P. G. (1883) 'Statistics of agricultural production', *Journal of the Royal Statistical Society*, 46: 1–58, reporting a survey by the *Farmer and Chamber of Agriculture Journal*.

1911: Acreages and livestock from agricultural statistics for Great Britain for 1911, part I, c. 6021, BPP 1912–13, CVI. Livestock units in 4.31–36 are: cow 1.0, other cattle 0.6, sheep 0.14, pigs 0.15. Yields are averages for 1906–15 from agricultural statistics for England and Wales for 1916, c. 8300, BPP 1917–18, XXXVI and agricultural statistics for Scotland for 1916, Part II, c. 170, BPP 1919, LI.

4.21 and 4.22 English data from Turner, M. E. (1980) *English Parliamentary Enclosure*, Folkstone, 180–1. More detail is in Tate W. E. (1978) *A Domesday of English Enclosure Acts and Awards*, (ed. Turner, M. E.), Reading. Welsh data courtesy of Dr J. Chapman, Department of Geography, Portsmouth Polytechnic.

4.23 and 4.24 1831 Census Enumeration Abstract, BPP 1833, XXXVI. 4.24 uses the method of Kussmaul, A. (1981), 170–1.

4.25–28, 4.38 Agricultural returns for Great Britain for 1895, c. 8073, BPP 1896, XCII.

4.37 Agricultural returns for Great Britain for 1896, c. 8502, BPP 1897, XCVIII.

4.49–4.52 Schedule A income tax assessments from Stamp, J. C. (1916), 54–6, using the method of Grigg, D. B. (1965) 'An index of regional change in English farming', *Transactions of the Institute of British Geographers*, 36: 55–67.

5 Rural settlements

See text for discussion of map sources.

6 Wages

For a full discussion of sources see Bowley (1898–1901), Bowley (1900), Gilboy (1934), Hunt (1973) and Hunt (1986).

7 Wind and water power

7.1 Sites plotted on a modern topographical map from C. and J. Greenwood's map of Somerset (London, 1822), C. Greenwood's map of Wiltshire (London, 1825) and A. Bryant's map of Gloucestershire (London, 1824). These maps were surveyed during the years 1819–23. Some watermills in Wiltshire marked on J. Andrews and A. Dury's map (London, 1773) and known to be working in 1820 but not shown by Greenwood have been added: see Rogers, K. (1976) *Wiltshire and Somerset Woollen Mills*, Edington, Wiltshire.

7.2 Sites plotted on a modern topographical map from T. Donald and T. Milne's map of Norfolk (London, 1797), J. Hodskinson's map of Suffolk (London, 1783) and J. Chapman and P. André's map of Essex (London, 1777). These maps were surveyed during the years 1772–1794.

7.3 Sites plotted on a modern topographical map from T. Jefferys' map of Yorkshire (London, 1771–2) surveyed 1767–70.

7.4 The sites shown in the preceding map have been enumerated by 10 km National Grid Squares.

8 Coal and steam power

The sources are fully discussed in the text. There is a wide variety of information in British Parliamentary Papers. See for instance the extensive list of coal prices at east and south coast English and Scottish ports in 1785, given in BPP 1871, XVIII, 1091, 1099 and 1160.

9 Transport

9.1 Adapted from Pawson (1977), 151.

9.2 Adapted from Freeman, M. J. and Longbotham, J. (1980) 'The Fordham Collection at the Royal Geographical Society: an introduction', *Geographical Journal*, 146, 226.

9.3 Adapted and extended from Freeman and Longbotham, op. cit., 228.

9.4 From information in Pigot & Co.'s *New Commercial Directory for 1828–9* (Manchester Central Library).

9.5 From West Yorkshire Record Office, County Records 51, Turnpike Trusts (6), Annual returns of turnpike trusts to Clerk of the Peace.

9.6 Adapted and extended from Dyos and Aldcroft (1969), 104–5.

9.7 Compiled from statistics in Langton, (1983), tables 1.1 and 1.2.

9.8 Adapted from Lewis (1970), 112–13.

9.9 Drawn from information supplied by S. R. Hughes, Royal Commission on Ancient Monuments in Wales (Swansea Valley Tramroads) and A. P. Wakelin (Neath Valley Tramroads).

9.10 Adapted from Pollins (1971), 32.

9.11 Adapted from Pollins (1971), 42.

9.12 Adapted from BPP, *Report of Joint Select Committee on Railway Companies' Amalgamation* (1872).

9.13 Adapted from *Bartholemew's Gazetteer of the British Isles* (London, 1904), plates 17–19.

9.14 As for 9.3.

9.15 Compiled from *Bradshaw's Railway Companion for August 1845* (London).

9.16 Compiled from *Bradshaw's August 1910 Railway Guide* (London, 1910).

9.17 Compiled from information in Nock, O. S. (1959) *The Railway Race to the North*, London.

9.18 From statistics in Carter, E. F. (1959), 448.

9.19 As for 9.13.

9.20 As for 9.13.

9.21 Adapted from Patmore (1961), 233.

9.22 From statistics in Ross (1904), 95.

9.23 From statistics in Hawke (1970), 170.

10 Sea trade

The Customs service has kept records of various aspects of the commodity composition, direction, volume and official value of trade since 1697, and of the real value of trade since the mid-nineteenth century. Unfortunately surviving records – especially for individual ports – differ from time to time. This is the principal reason why the sample years vary slightly according to the subject.

For 1772 to 1808 the Public Record Office Customs 17 series has a reasonable analysis of trade and shipping movements nationally, and to some extent by ports, and of ships 'belonging' before 1786 and registered from 1788. There is no detailed information for the earliest period of industrialization, but the material for 1791 is particularly good, and provides an excellent benchmark for later comparisons. From 1808 these Customs reports are available as Trade and Navigation Accounts (later Annual Statements of Trade and Navigation) in British Parliamentary Papers. The year 1841

– BPP 1842 (409) XXXIX – was chosen in preference to 1851 because it antedates the impact of railways, steamships and new docks. Shipping movement figures for 1900 may be found in BPP 1901 (604) LXXV, and registered shipping in BPP 1902 (1113) C.

Indications of changing composition and sources of imports are best gained by sampling statistics transcribed in Schumpeter, E. B. (1960) *English Overseas Trade Statistics, 1697–1808*, Oxford, and the annual accounts and statements in BPP, which become very detailed and more useful in the later nineteenth century, often providing 5-year comparisons.

So far as overall trade is concerned, volume is easily found in Customs 17 and BPP, but the official values were not adjusted with price changes during the industrial revolution. The value of trade in these tables and maps is based on the reworking of raw trade figures for 1794–6 and 1844–6 by Ralph Davis (1979), and on the real values contained in the Annual Statement of Trade for 1900.

10.16 is based upon Jackson (1983b), 33 and 45; **10.19** on Newham (1913), 64; and **10.19** insert on Jackson (1972), 447.

11 Textiles

11.1–2 Data compiled from Hatley, V. A. (ed.) (1973) *Northamptonshire Militia Lists 1777*, Nottinghamshire Record Society, vol. 25. Detached parts of hundreds have been united with surrounding hundreds and the data adjusted accordingly.

11.3–4 Data compiled from *The Craven Muster Roll 1803*, North Yorkshire County Record Office Publication no. 9, Northallerton 1976. See alternative analyses in Lawton, R. (1954) 'The economic Geography of Craven in the early 19th century', *Transactions of the Institute of British Geographers*, 20: 93–111. 'Others' chiefly includes wool combers and cotton spinners (in factories and workshops).

11.5 Based on figs 1 and 2 and appendix I of Jenkins (1975).

11.6 Data from Daniels, G. W. (1930) 'Samuel Crompton's census of the cotton industry in 1811', *Economic History*, 2: 108–9.

11.7 Based upon the Ordnance Survey 1:1056 (60 inches to one mile) plans of Manchester and Salford, surveyed 1844–9, and the 1:10560 map surveyed at the same time.

11.8 Data compiled from the 1861 *Census Abstract*, vol. 2, pt 2, occupation tables, BPP (HC) (3221) LIII, pt 2, 223–468, *passim*. Those aged 20 and over in the manufacture and finishing of yarn, cloth, ribbon and hosiery are expressed as a percentage of all those aged 20 and over except females in order IV. The areas are registration districts; the only other districts exceeding 20 per cent on this criterion were Carlisle, Melksham and Stroud.

11.9 Data compiled from BPP (HC) 1839 XLIII.

11.10–13 Data compiled from the 1861 *Census Abstract*, vol. 2, occupation tables: BPP (HC) 1863 (3221) LIII. In England and Wales data for registration districts; for Scotland registration counties.

11.14 Data from John Worrall and Co. (1890) *The Cotton Spinners' and Manufacturers' Directory*, Oldham. Each district includes mills in both its town and adjacent districts.

11.15 Data calculated from John Worrall and Co. (1910) *The Cotton Spinners' and Manufacturers' Directory*, Oldham. The larger the circle or square, the greater the swing to spinning or weaving respectively. Several districts 'swung' against their specialism because changing machine efficiency and fixed capital investment increased the spindle/loom ratio for the whole region.

11.16–17 Data compiled from the 1911 *Census Abstract*, vol. 10, occupation and industry tables: BPP (HC) (Cd. 7660) 1914–16, LXXXI. The areas mapped here housed 96 per cent of British cotton workers, 61 per cent of whom were females.

12 Chemicals

Like the industry itself, sources on chemicals are diverse. For raw materials, there is information about salt (BPP 1836, XVII); the Mining Records of the Geological Survey (and later *Mineral Statistics of the UK*) provide some data on other minerals. BPPs contain the valuable reports and minutes of evidence of the committees on noxious vapours of 1862 and 1878. From the mid-1860s the BPPs contain the valuable annual Alkali Inspector's report. The 1882 list of chemical works (BPP 1882 LVII) is especially useful. The sources listed, as well as trade journals or directories, or the 1890 *Prospectus of the United Alkali Company*, do not usually provide quantitative data.

13 Brewing and distilling

13.1–4 1788/99: *Two Reports from the Select Committee on the Distillery*, BPP 1803 (1st series) XI, appendix B2, 200–1.
1833: *The Seventh Report of the Commissioners of Inquiry into the Excise Establishment*, BPP 1834, XXV, appendix 67, 229–32.
1900: Wilson (1940), appendix F, tables 3, 8, and 9.

13.5–8 1754/5: Mathias, P. (1959) *The Brewing Industry in England 1700–1830*, Cambridge, 540. We are grateful to Professor Mathias for permission to use this map in compiling 13.5.
1822: Account of the number of barrels on which duty has been paid, BPP 1822, XXI.
1850: *Accounts and Papers, Licensed breweries . . .* , BPP 1850, LII.
1900: *Accounts and Papers, Licensed breweries . . .* , BPP 1900, LXXVII.

14 Leather footwear

The data were taken from the printed Census abstract in British Parliamentary Papers. For a full discussion of this source see articles by Anderson, Tillott and Armstrong, in Wrigley, E. A. (ed.) (1972) Nineteenth Century Society: *Essays in the Use of Quantitative Methods for the Study of Social Data*, Cambridge. In **14.3** registration counties are used and data are from enumerators'manuscript schedules of the 1851 Census. The data for the inset in **14.4** are taken from the local studies listed in the bibliography. The main map is compiled from data in Butnam (1911) *Shoe and Leather Trade of the UK*, London.

15 Iron and steel

The sources are fully discussed in Riden (1977) and Riden (1980).

16 Shipbuilding

16.1–3 Compiled from data in the printed Census abstract in BPP.
16.4 Compiled from data in *The Annual Statement of the Navigation and Shipping of the UK for 1911*, vol. 85, 314–17.
16.5–7 From *An Account of the Number of Vessels and their Tonnage that have been Built at the Ports of Great Britain and Ireland*, BPP (HC) 1826–27, XVIII. Also from *Returns of the Number and Tonnage of Vessels the Building of which was Completed in the Year 1871 at each Port in the United Kingdom*, BPP (HC) 1872, LIII.

17 Engineering

Census of Great Britain, 1851, Population Tables II, Volumes 1 and 2, London, 1854.
Census of Scotland, 1911: Report on the Twelfth Decennial Census of Scotland, Volume 1, Parts 1 to 37, Edinburgh, 1912.
Census of England and Wales, 1911, Volume X, Occupations and Industries, Part 2, London, 1914.

18 Services

The compilation of data for the maps is fully discussed in Lee (1979).

19 Banking and finance

Monetary and financial conditions were a particular concern of Parliament during the century after 1775, with the consequence for historians that British Parliamentary Papers, both blue books and accounts, are a rich source for those interested in developments in banking. Accordingly most of the maps in chapter 19 are based upon BPP with, for instance, the county by county distribution of issuing licences coming from data contained in BPP 1822, XXI, 67, the total number of licences granted being found in BPP 1831–2, VI, 599, and the number of private and joint stock banks being given by BPP 1843, LII, 9. Similarly data for the branches of the Bank of England came from BPP 1837–8, VII, 121, 124, while the data for the private note issue was drawn from Appendix No. 24 to the Select Committee on Banks of Issue (1841). Other sources employed have been bank balance sheets as utilized by Collins (1983) and Capie and Webber (1985), and directories, drawing from the work of Pressnell (1956), Munn (1981) and Nishimura (1971).

20 Wealth and the wealthy

The sources are fully discussed in the text.

21 Poor law and pauperism

Report by a Committee of the General Assembly on the Management of the Poor in Scotland, BPP 1839, XX.
A Return Showing the Number of Paupers in England and Wales, 1839 and 1840, BPP 1841, XXI.
The 24th Annual Report of the Board of Supervision, C.5, BPP 1870, XXXV.
The 21st Annual Report of the Poor Law Board, BPP 1868–9, XXVIII.
The 22nd Annual Report of the Poor Law Board, C.123, BPP 1870, XXXV.
The 15th Annual Report of the Local Government Board (Scotland), Cd 5228, BPP 1910 XL.
The 38th Annual Report of the Local Government Board (England), Cd 4786, BPP 1909, LXXV.
Statistical Memorandum and Charts prepared in the Local Government Board relating to Public Health and Social Conditions, BPP 1909, CIII.

22 Urbanization

22.1–6 Derived from the data set used by Robson (1973). I am grateful to Professor Robson for allowing me to use this data and supplying me with a machine-readable copy. This, together with a package which attached grid references to the town names (GAZ by Mr John Welford of CAST, Edinburgh University) and the mapping package GIMMS (Mr Tom Waugh, Department of Geography, Edinburgh University) enabled me to compile these maps.

22.7 From Cannadine (1980) and Ordnance Survey map 1:10560, 2nd edn, 1900, Sussex (East) Sheet 80 SW.

22.8 Derived from Ordnance Survey 1:10560, Yorkshire Sheet 6, surveyed 1853 and published 1857.

22.9 Derived from Ordnance Survey 1:10560, Yorkshire (North Riding) Sheet 6. SW and SE divisions, surveyed 1892–3 and published 1895.

22.10–11 From Checkland (1964), 44 and 47. My thanks to Professor Checkland for permission to use his work.

22.12 From *A New Map of the University and City of Oxford*, 1812, published by R. Pearse, High Street, Oxford.

22.13–14 From the map and drawing prepared by Barbara Morris for Morris (1971).

22.15 From H. W. Acland (1856) *Memoir of the Cholera at Oxford in the year 1854*, London.

22.16 From Business Archives, Lupton 127, Brotherton Library, University of Leeds. My thanks to Mr P. Morrish for showing me these papers.

22.17 Derived from Ordnance Survey 1:1056 Leeds, Sheet 11, surveyed 1847 and published 1850.

22.18–19 *First Report of the Commissioners for Inquiring into the State of Large Towns and Populous Districts*, BPP (HC) 1844, XVII, appendix, 143.

22.20–21 *Report of an Enquiry by the Board of Trade into Working Class Rents, Housing and Retail Prices ... in the Principal Industrial Towns of the United Kingdom*, 590–3. BPP (HC) 1908, CVII.

22.22–23 *Returns from Gas Companies Established by Act of Parliament*, BPP (HC) 1847, XLIV.

22.24–25 *Return of Street and Road Tramways Showing the Amount of Capital Authorized and the Length Opened for Public Conveyance from the Passing of 'The Tramways Act, 1870' down to the 30th June 1876*, BPP (HC) 1877, LXXIII.

22.26 Derived from the Ordnance Survey five-foot plan of Edinburgh, Sheets 35 and 36, published 1852, and five-foot plan of Edinburgh, sheets 111.8.16 and 21, re-surveyed 1893–4 and published 1895.

22.27 *Royal Commission on London Traffic*, vol. 5, BPP (HC) 1906, XLIV, plate XVIII.

22.28 *Royal Commission on London Traffic*, vol. 7, BPP (HC) 1906, XLII, appendix J, diagram 12.

22.29 As for 22.20–21 but pp. 6–7.

22.30 Derived from Ward (1980), 152. My thanks to Professor Ward for permission to use this material.

22.31 From Pooley (1977). My thanks to Colin Pooley for permission to use his work.

22.32 From the *Census Abstract* of 1841 and *The Leeds Parliamentary Poll Book*, Leeds 1841.

23 Retail patterns

These are discussed in Alexander (1970) and Shaw and Wild (1979). The figures quoted in the text are for the following selected towns: Leeds, Hull, Halifax, Huddersfield, Rochdale, Oldham, Wakefield, Lancaster, York and Beverley.

24 Labour protest 1780–1850

24.1 After Map 27 in Charlesworth (1983).

24.2 Dobson (1980) appendix.

24.3 Thomas, M. I. (1970) *The Luddites*, Newton Abbot, Appendix; Darvall, F. O. (1934) *Popular Disturbances and Public Order* *in Regency England*, London, chs 4–6; Dinwiddy, J. (1979) 'Luddism and politics in the northern counties', *Social History*, 4 (1), 33–63.

24.4 Reproduced by kind permission of Croom-Helm from Charlesworth (1983).

24.5 Jones, D. (1975) *Chartism and the Chartists*, London, ch. 9; Wilson, A. (1970) *The Chartist Movement in Scotland*, Manchester, 84–5, 273. I would like to thank Dr Wilson for his help in clarifying a number of points.

24.6 Mather, F. C. (1974) 'The General Strike of 1842' in Quinault, R. and Stevenson, J. (eds) *Popular Protest and Public Order*, London, 135–6.

25 Unionization

The source for **25.1** is the index to the collection of friendly society rules created by the Registrar of Friendly Societies (Public Record Office, holding F.S.1, index in F.S.2); trade societies were identified solely by name, and the industrial classification used was that given in Armstrong, W. A. (1972) 'The use of information about occupation', in Wrigley, E. A. (ed.) *Nineteenth Century Society*, Cambridge, 226–310. The immediate sources for **25.2–5** are: Fyrth, H. J. and Collins, H. (1959) *The Foundry Workers*, Manchester, 21 (Ironfounders); Mechanics' Friendly Institution, *Rules*, 1824 (copy in Manchester Working Class Movement Library) (Mechanics); Newcastle and Country United Tanners, *Rules*, 1826 (British Library 8275.bb.4); Kiddier, W. (1931) *The Old Trade Unions*, London, 16–16 (Brushmakers). **25.6** uses information tabulated in Webb, S. and B. (1894) *The History of Trade Unionism*, 1st edn, London, 489–91; it proved impossible to locate the underlying calculations in their papers, hence the only adjustment possible was to state union members as a fraction of employed rather than total population, using statistics given in Lee, (1979), 397, with employment statistics again from Lee (1979).

26 Popular institutions

All sources are from British Parliamentary Papers unless otherwise stated.

26.1 *Return of the Number of Friendly Societies, filed by the Clerks of the Peace of each County, Riding and Division, City, Borough, and Place in Great Britain and Ireland, since 1st January 1793 to the Time of Making such Returns*, BPP 1831–2, XXVI, 291, and *Registry of Friendly Societies. Index to Rules and Amendments Series 1, 1784–1875*, Public Record Office, FS 2/1–13.

26.2 *Reports of the Registrar of Friendly Societies in England for the Year ending 31st December 1872*, BPP 1873, LXI, 87, Appendix, and similar *Return from the Registrar of Friendly Societies in Scotland*.

26.3 Newspaper press and Public Record Office FS 2/1–13.

26.4 *Abstract of Returns Furnished to the Registrar of Friendly Societies in England by Industrial and Provident (Co-operative) Societies for the Year ending 31st December 1872*, BPP 1873, LXI, 349, and similar *Return from the Registrar of Friendly Societies in Scotland*.

26.5 *Return of Licenses granted by the Board of Inland Revenue for the Years ending March 1871 and 1872*, BPP 1872, XXXVI, 301; *Excise Licenses (Scotland). Return of the Number of Inns and Hotels, Public Houses, and other Premises in which Spirits, Wines and other Exciseable Liquor, and Table Beer, is Retailed in each County, City and Burgh in Scotland . . . at Whitsunday 1841, 1851, 1861, 1871 and 1881*, BPP 1884, XLVII, 213.

26.6 *Alliance News*, 1872.

26.7 Co-operative Union *Annual Congress Reports* 1884–99, Manchester; Co-operative Union *Co-operative Directory*, 1900, Manchester.

26.8 *Liquor Licensing Laws (Royal Commission). Return of Clubs in Great Britain and Ireland*, BPP 1898, XXXVII, 1.

27 Sport

These are discussed in Vamplew (1976) and Vamplew (1982). I am grateful for the valuable assistance of Andrew Little, Senior Cartographer, Flinders University, South Australia.

28 Languages and dialects

28.1 Withers (1984), 47, 81 and 180.
28.2 Withers (1984), 228 and 231.
28.3 Pryce (1978a), 11, 16, 20 and 31.
28.4 Pryce (1978a), 23 and 31.
28.5 Jenner (1875), 175.
28.6 Graph and county symbols were derived from a count of the items listed in Skeat and Nodal (1877) including the addenda of Part III. Graphic extrapolation to 1880 was calculated by dividing the recorded number by 7 and multiplying by 10. All editions of single works were counted. All undated publications were included on the map. On the graph, undated publications were allocated to decades in which the same authors issued dated publications if any of the latter existed. Editions of the *Song of Solomon* were taken from Skeat and Nodal (1877) and Wright (1896).
28.7 Data taken from Vicinus (1974), 233n. They derive from Waugh's unpublished journals, which are kept in Manchester Central Library.

29 Education

29.1 From *Select Committee Appointed to Inquire into the Education of the Lower Orders (England and Wales)*, BPP 1818, VI.
29.2-3 *Education Census, 1851*, BPP 1852-3, XC. The child population used in **29.1-3** has been taken as one-sixth of the total population. This was a notional figure widely used in the mid- and late nineteenth century and provides a useful basis for comparison. The rationale for its use was given by Horace Mann in his introduction to the 1851 Education Census, but see 2.13-20 on the regional variations in age structure.
29.4-6 *London School Board Statistical Reports*.
29.7 Booth (1891).
29.8 Marsden (1977c).

30 Religion

30.1-3 *Religious Census of England and Wales*, BPP 1852-3, LXXXIX, and *Religious Census of Scotland*, BPP 1854, LIX.
30.4-7 The Methodist statistics were taken from the Minutes of Conference of the various denominations.
30.8-11 Mudie Smith, R. (1904) *The Religious Life of London*, and *London Statistics 1914-15*; see McLeod (1974).

31 The electoral system

Dod, C. R., *Electoral Facts, 1832-1853*. Reprint ed. Hanham, H. J.; McCalmont, F. H., *Parliamentary Poll Book, 1832-1910*. 8th edn, ed. Vincent, J. and Stenton, M., Brighton, 1971; Porritt (1903).

Bibliography

Aikin, A. J. and McArthur, T. (1979) *Languages of Scotland*, Edinburgh.

Albert, W. (1972) *The Turnpike Road System in England 1663–1840*, Cambridge.

Aldcroft, D. H. and Freeman, M. J. (eds) (1983) *Transport in the Industrial Revolution*, Manchester.

Alexander, D. (1970) *Retailing in England During the Industrial Revolution*, London.

Anon. (1963) *A Century of Agricultural Statistics: Great Britain 1866–1966*, London.

Ardener, S. (1981) *Women in Space*, London.

Ashplant, T. G. (1981) 'London Working Men's Clubs, 1875–1914', in Yeo, E. and Yeo, S. *Explorations in the History of Labour and Leisure*, Brighton, 241–70.

Ashton, T. S. (1945) 'The bill of exchange and private banks in Lancashire 1790–1830', *Economic History Review*, 15.

Aspinall, A. (1949) *The Early English Trade Unions: Documents from the Home Office Papers in the Public Record Office*, London.

Attfield, J. (1981) *With Light of Knowledge: A Hundred Years of Education in the Royal Arsenal Co-operative Society, 1877–1977*, London.

Bagwell, P. (1983) 'Coastal shipping', in Aldcroft, D. H. and Freeman, M. (eds) *Transport in the Industrial Revolution*, Manchester.

Baker, W. J. (1983) 'The state of British sport history', *Journal of Sport History*, 10 (1), 53–66.

Bale, J. (1982) *Sport and Place*, London.

Banbury, P. (1971) *Shipbuilders of the Thames and Medway*, Newton Abbot.

Barnes, F. A. (1970) 'Settlement and landscape changes in a Caernarvonshire slate quarrying parish', in Osborne, R. H., Barnes, F. A. and Doornkamp, J. C. (eds), *Geographical Essays in Honour of Professor K. C. Edwards*, Nottingham, 119–30.

Bateman, J. (1876) *The Great Landowners of Great Britain and Ireland*, London, reprinted Leicester, 1971.

Beckett, J. V. (1981) *Coal and Tobacco: The Lowthers and the Economic Development of West Cumberland, 1660–1760*, Cambridge.

Berg, M (1985) *The Age of Manufactures, 1700–1820*, London.

Bilham, E. G. (1938) *The Climate of the British Isles*, London.

Birch, A. (1967) *The Economic History of the British Iron and Steel Industry 1784–1879: Essays in Industrial and Economic History with Special Reference to the Development of Technology*, London.

Bird, J. (1963) *The Major Seaports of the United Kingdom*, London.

Booth, C. (ed.) (1891) *Life and Labour of the People*: vol. 2, London, 477–588.

Bowley, A. L. (1898–1901) 'The statistics of wages in the United Kingdom during the last hundred years: Parts I and II, agricultural wages; Parts VI, VII and VIII, wages in the building trades', *Journal of the Royal Statistical Society*, LXI–LXIV.

Bowley, A. L. (1900) *Waghes in the United Kingdom in the Nineteenth Century*, Cambridge.

BPP (1874) *Fourth Report of the Commissioners Appointed to Inquire into Friendly and Building Societies*. Appendix IV: Some statistics of Friendly Societies in England and Wales, with special reference to those of Lancashire, BPP 1874: XXIII, pt 1, 295.

BPP (1874) *Fourth Report of the Commissioners appointed to Inquire into Friendly and Building Societies*. Reports of the Assistant Commissioners, Scotland with Cumberland, Durham, Northumberland and Westmoreland, BPP 1874: XXIII, pt 2, 627.

Brookes, C. (1978) *English Cricket*, Newton Abbot.

Brundage, A. (1978) *The Making of the New Poor Law*, London.

Cage, R. (1981) *The Scottish Poor Law 1745–1845*, Edinburgh.

Calhoun, C. (1982) *The Question of Class Struggle*, Oxford.

Cannadine, D. (1980) *Lords and Landlords: The Aristocracy and the Towns, 1774–1967*, Leicester.

Cannon, J. (1973) *Parliamentary Reform,*

1640–1832, Cambridge.

Capie, F. and Webber, A. (1985) *A Monetary History of the United Kingdon, 1870–1982*, London.

Carter, E. F. (1959) *An Historical Geography of the Railways of the British Isles*, London.

Carter, H. *et al.* (1974) *An Advanced Geography of the British Isles*, Amersham.

Chambers, J. D. (1953) 'Enclosure and labour supply in the industrial revolution', *Economic History Review*, 2nd series, 5: 319–43.

Chambers, J. D. and Mingay, G. E. (1966) *The Agricultural Revolution, 1750–1880*, London.

Chandler, T. J. and Gregory, S. (1976) *The Climate of the British Isles*, London.

Chapman, S. D. (1967) *The Early Factory Masters*, Newton Abbot.

Chapman, S. J. (1904) *The Lancashire Cotton Industry*, Manchester.

Charlesworth, A. (ed.) (1983) *An Atlas of Rural Protest in Britain 1548–1900*, Beckenham.

Checkland, S. G. (1964) 'The British industrial city as history: The Glasgow case', *Urban Studies*, 1.

Checkland, S. G. (1975) *Scottish Banking: A History*, Glasgow.

Chorley, G. P. H. (1981) 'The agricultural revolution in northern Europe, 1750–1880: nitrogen, legumes and crop producitivity', *Economic History Review*, 2nd series, 34: 71–93.

Church, R. (1966) 'Messrs Gotch and Sons and the rise of the Kettering footwear industry', *Business History*, 8: 140–9.

Clapham, Sir J. H. (1944) *The Bank of England: A History*, 2 vols, Cambridge.

Clarke, P. F. (1971) *Lancashire and the New Liberalism*, Cambridge.

Clegg, H. A., Fox, A. and Thompson, A. F. (1964) *A History of British Trade Unions since 1889*, vol. 1, Oxford.

Clemenson, H. A. (1982) *English Country Houses and Great Estates*, London.

Cole, G. D. H. (1944) *A Century of Co-operation*, Manchester.

Collins, E. J. T. (1981) 'The age of machinery', in Mingay, G. E. (ed.) *The Victorian Countryside*, London, 200–13.

Collins, M. (1972) 'The Bank of England at Liverpool, 1827–1844', *Business History*, 14.

Collins, M. (1983) 'The long-term growth of the English banking sector and money stock, 1844–80', *Economic History Review*, 2nd series, 36.

Cope, S. R. (1942) 'The Goldsmiths and the development of the London money market during the Napoleonic Wars', *Economica*, n.s., 9.

Coppock, J. T. (1971) *An Agricultural Geography of Great Britain*, London.

Coppock, J. T. (1976a) *An Agricultural Atlas of Great Britain*, London.

Coppock, J. T. (1976b) *An Agricultural Atlas of Scotland*, Edinburgh.

Coppock, J. T. (1984) 'Mapping the agricultural returns: a neglected tool of historical geography', in Reed, M. (ed.) *Discovering Past Landscapes*, London, 8–55.

Corfield, P. (1982) *The Impact of English Towns 1700–1800*, Oxford.

Cornford, J. (1963) 'The transformation of conservatism in the late 19th century', *Victorian Studies*, 7.

Crafts, N. F. R. (1985) *British Economic Growth During the Industrial Revolution*, Oxford.

Crick, W. F. and Wadsworth, J. E. (1936) *A Hundred Years of Joint Stock Banking*, London.

Cronkshaw, P. (1948) 'The Rossendale Valley: an industrial romance', *Transactions of the Lancashire and Cheshire Antiquarian Society*, 60: 29–35.

Crowther, M. A. (1981) *The Workhouse System, 1834–1929*, London.

Curtis, L. F., Courtney, F. M. and Trudgill, S. T. (1976) *Soils in the British Isles*, London.

Dahlman, C. (1980) *The Open Field System and Beyond*, Cambridge.

Darley, G. (1978) *Villages of Vision*, London.

Daunton, M. (1977) *Coal Metropolis: Cardiff, 1870–1914*, Leicester.

Daunton, M. (1983) *House and Home in the Victorian City: Working Class Housing, 1850–1914*, London.

Davis, R. (1954) 'English foreign trade 1660–1700', *Economic History Review*, 2nd series, 7.

Davis, R. (1962a) *The Rise of the English Shipping Industry in the Seventeenth and Eighteenth Centuries*, London.

Davis, R. (1962b) 'English foreign trade, 1700–70', *Economic History Review*, 2nd series, 15.

Davis, R. (1973) *The Rise of the Atlantic Economies*, London.

Davis, R. (1979) *The Industrial Revolution and British Overseas Trade*, Leicester.

Deane, P. and Cole, W. A. (1962) *British Economic Growth, 1688–1959*. Cambridge, 2nd edn 1969.

Devine, T. (1975) *The Tobacco Lords*, Edinburgh.

Devine, T. M. (1975) 'The rise and fall of illicit whisky distilling in Northern Scotland, c.1780–1840', *Scottish Historical Review*, 54.

Digby, A. (1981) *Pauper Palaces*, London.

Dixon, P. J. (1979) 'School attendance in Preston: some socio-economic influences', in Lowe, R. (ed.), *New Approaches to the Study of Popular Education 1851–1902*, Leicester, 43–58.

Dixon, P. J. (1982) 'The lower middle class child in the grammar school: a Lancashire industrial town, 1850–1875', in Searby, P. (ed.) *Educating the Victorian Middle Class*, Leicester.

Dobson, C. R. (1980) *Masters and Journeymen: A Prehistory of Industrial Relations 1717–1800*, London.

Dod, C. R. (1853) *Electoral Facts, from 1832 to 1853, Impartially Stated*, London, reprint, Hanham, H. J. (ed.) (1972), London.

Dodgshon, R. A. (1980) 'Medieval settlement and colonisation', in Parry, M. L. and Slater, T. R. (eds) *The Making of the Scottish Countryside*, London, 45–69.

Dodgshon, R. A. (1981) *Land and Society in Early Scotland*, Oxford.

Donnachie, I. (1979) *A History of the Brewing Industry in Scotland*, Edinburgh.

Doornkamp, J. C. and Gregory, S. (eds) (1980) *Atlas of Drought in Britain*, London.

Dow, G. (1959) *Great Central*, 2 vols, London.

Drake, M. (1972) 'The census 1801–91', in Wrigley, E. A. (ed.) *Nineteenth Century Society*, Cambridge.

Duckham, B. F. (1967) *The Yorkshire Ouse*, Newton Abbot.

Dunford, M. and Perrons, D. (1983) *The Arena of Capital*, London.

Dunning, E. and Sheard, K. (1979) *Barbarians, Gentlemen and Players*, Oxford.

Durham County Council (1951) *County Development Plan, 1951, Written Analysis*, Durham.

Durkacz, E. (1983) *The Decline of the Celtic Languages*, Glasgow.

Dyos, H. J. and Aldcroft, D. H. (1969) *British Transport: An Economic Survey from the Seventeenth Century to the Twentieth*, Leicester.

Dyos, H. J. and Wolff, M. (eds) (1973) *The Victorian City: Images and Realities*, London.

Eastwood, T. (1964) *Stanford's Geological Atlas of Great Britain*, London.

Everitt, A. (1979) 'Country, county and town: patterns of regional evolution in England', *Transactions of the Royal Historical Society*, 5th series, 29.

Fairlie, S. (1969) 'The corn laws and British wheat production 1829–1876', *Economic History Review*, 2nd series, 22: 88–116.

Falkus, M. E. (1967) 'The British gas industry before 1850', *Economic History Review*, 20.

Farnie, D. A. (1979) *The English Cotton Industry and the World Market, 1815–1896*, Oxford.

Farnie, D. A. (1980) *The Manchester Ship Canal and the Rise of the Port of Manchester, 1894–1975*, Manchester.

Flinn, M. W. (ed.) (1977) *Scottish Population History*, Cambridge.

Flinn, M. W. and Stoker, D. (1984) *The History of the British Coal Industry, vol. 2: 1700–1830, The Industrial Revolution*, Oxford.

Floud, R. and McCloskey, D. (eds) (1981) *The Economic History of Britain since 1700*, Cambridge, vol. I, chs 4 and 10; vol. II, ch. 8.

Floud, R. C. (1976) *The British Machine Tool Industry, 1850–1914*, Cambridge.

Fong, H. D. (1930) *The Triumph of the Factory System in England*, Tientsin, China.

Foster, H. J. (1982) 'Private, proprietary and public elementary schools in a Lancashire residential town: a contest for the patronage of the lower middle classes 1870–1900', in Searby, P. (ed.) *Educating the Victorian Middle Class*, Leicester.

Fox, C. (1959) *The Personality of Britain*, Cardiff.

Fraser, D. (1984) *The Evolution of the British Welfare State*, London.

Fraser, D. (ed.) (1976) *The New Poor Law in the 19th Century*, London.

Freeman, M. J. (1980) 'Road transport in the English industrial revolution: an interim reassessment', *Journal of Historical Geography*, 6: 17–28.

Fulford, R. (1953) *Glyns 1753–1953*, London.

Gaffin, J. and Thomas, D. (1983) *Caring and Sharing: the Centenary History of the Co-operative Women's Guild*, Manchester.

Garnett, R. G. (1972) *Co-operation and the Owenite Socialist Communities in Britain, 1825–45*, Manchester.

Gash, N. (1952) *Politics in the Age of Peel*, London.

Gay, J. D. (1971) *The Geography of Religion in England*, London.

Gilbert, A. D. (1976) *Religion and Society in Industrial England 1740–1914*, London.

Gilboy, E. W. (1934) *Wages in Eighteenth-Century England*, Cambridge, Mass.

Glass, D. V. (1973) *Numbering the People*, Farnborough.

Goodhart, C. A. E. (1972) *The Business of Banking 1891–1914*, London.

Gosden, P. H. J. H. (1961) *The Friendly Societies in England, 1815–1875*, Manchester.

Gourvish, T. R. (1980) *Railways and the British Economy 1830–1914*, London.

Greenall, R. L. (1975) 'The rise of industrial Kettering, Northamptonshire', *Past and Present*, 5 (3): 253–66.

Gregory, D. (1982) *Regional Transformation and the Industrial Revolution: A Geography of the Yorkshire Woollen Industry*, London.

Gregory, Sir T. E. (1929) *Select Statutes, Documents and Reports Relating to British Banking 1832–1923*, London, 2 vols.

Gregory, Sir T. E. (1936) *The Westminster Bank through a Century*, London, 2 vols.

Grigg, D. B. (1967) 'The changing agricultural geography of England: a commentary on the sources available for the reconstruction of the agricultural geography of England, 1770–1850', *Transactions of the Institute of British Geographers*, 41: 73–96.

Hall, B. T. (1922) *Our Sixty Years: The Story of the Working Men's Club and Institute Union*, London.

Hammond, J. L. and Hammond, B. (1979) *The Skilled Labourer*, reprint, London.

Hanham, H. J. (1959) *Elections and Party Management: Politics in the Time of Disraeli and Gladstone*, London.

Hardie, D. W. F. and Pratt, J. D. (1966) *A History of the Modern British Chemical Industry*, Oxford.

Harley, J. B. (1964) *The Historian's Guide to Ordnance Survey Maps*, The Standing Conference for Local History, the National Council for Social Service, London.

Harley, J. B. (1972) *Maps for the Local Historian*, Standing Conference for Local History, National Council for Social Service, London.

Harris, A. (1961) *The Rural Landscape of the East Riding of Yorkshire*, Oxford.

Harrison, B. (1971) *Drink and the Victorians: The Temperance Question in England 1815–1872*, London.

Harrison, B. (1973) 'Pubs', in Dyos, H. J. and Wolff, M. (eds) *The Victorian City: Images and Realities*, London, vol. I, 161–90.

Harvey, D. (1982) *The Limits to Capital*, Oxford.

Hawke, G. (1970) *Railways and Economic Growth in England and Wales 1840–1914*, Oxford.

Hechter, M. (1975) *Internal Colonialism*, London.

Hills, R. L. (1978) *Power in the Industrial Revolution*, Manchester.

Hobsbawm, E. J. (1964) *Labouring Men*, London.

Hobsbawm, E. J. (1984) *Worlds of Labour*, London.

Hodgen, M. T. (1952) *Change and History*, New York.

Hohenburg, P. M. (1967) *Chemicals in Western Europe 1850–1914*, Chicago.

Hoole, K. (1972) *The Hull and Barnsley Railway*, Newton Abbot.

Hope-Jones, A. (1939) *Income Tax in the Napoleonic Wars*, Cambridge.

Howell, D. W. (1977) *Land and People in Nineteenth-Century Wales*, London.

Hudson, K. (1968) *Towards Precision Shoemaking: C. & J. Clark Ltd and the Development of the British Shoe Industry*, Newton Abbot.

Hughes, J. R. T. and Reiter, S. (1958) 'The first 1,945 British steamships', *Journal of the American Statistical Association*, 53.

Hull, E. (1905) *The Coal-Fields of Great Britain*, London.

Hunt, E. H. (1973) *Regional Wage Variations in Britain, 1850–1914,* Oxford.

Hunt, E. H. (1986) 'Industrialization and regional inequality: wages in Britain, 1760–1914', *Journal of Economic History*, 46.

Hurst, M. C. (forthcoming) *Countries, provinces and constituencies: The United Kingdom at politics, 1830–1918.*

Hyde, C. K. (1977) *Technological Change and the British Iron Industry, 1700–1870*, Princeton, NJ.

Hyde, F. E. (1971) *Liverpool and the Mersey*, Newton Abbot.

Irving, R. J. (1976) *The North Eastern Railway Company, 1870–1914*, Leicester.

Jackman, W. T. (1966) *The Development of Transportation in Modern England*, London.

Jackson, G. (1971) *Grimsby and the Haven Company, 1796–1846*, Grimsby.

Jackson, G. (1972) *Hull in the Eighteenth Century*, Oxford.

Jackson, G. (1983a) 'The Ports', in Aldcroft, D. H. and Freeman, M. (eds) *Transport in the Industrial Revolution*, Manchester.

Jackson, G. (1983b) *The History and Archaeology of Ports*, London.

Jeffreys, J. B. (1954) *Retail Trading in Britain 1850–1950*, Cambridge.

Jenkins, D. T. (1973) 'Early factory development in the West Riding of Yorkshire', in Harte, N. B. and Ponting, K. (eds) *Textile History and Economic History: Essays in Honour of Miss Julia de Lacy Mann*, Manchester.

Jenkins, D. T. (1975) *The West Riding Wool Textile Industry, 1770–1835: A Study in Fixed Capital Formation*, Edington, Wiltshire.

Jenner, H. (1875) 'The Manx Language: its grammar, literature and present state', *Transactions of the Philological Society*.

Jewell, A. (ed.) (1975) *Victorian Farming: A Sourcebook*, Winchester.

Jones, B. (1894) *Co-operative Production*, 2 vols, Oxford.

Jones, E. L. (1968) *The Development of English Agriculture, 1815–1873*, London.

Jones, E. L. (1974) *Agriculture and the Industrial Revolution*, Oxford.

Jones, H. (1983) 'Population patterns and processes from *c.*1600', in Whittington, G. and Whyte, I.D. *An Historical Geography of Scotland*, London.

Jones, I. G. (1981) *Explanations and Exploration: Essays in the Social History of Wales*, Gwasg Gomer.

Jones, P. N. (1969) *Colliery Settlement in the South Wales Coalfield 1850 to 1926*, University of Hull, Occasional Papers in Geography, 14.

Joslin, D. M. (1954) 'London private bankers, 1720–1785', *Economic History Review*, 9.

Joyce, P. (1980) *Work, Society and Politics*, Brighton.

Kain, R.J.P. and Prince, H.L. (1985) *The Tithe Surveys of England and Wales*, Cambridge.

Kanefsky, J. W. (1979) 'Motive power in British industry: the accuracy of the 1870

Factory Return', *Economic History Review*, 2nd series, 32.

Kanefsky, J. W. and Robey, J. (1980) 'Steam engines in eighteenth-century Britain: A quantitative assessment', *Technology and Culture*, 21: 2.

Kenwood, A. G. (1965) 'Port investment in England and Wales, 1851–1913', *Yorkshire Bulletin of Economic and Social Research*, 17.

Killick, J. R. and Thomas, W. A. (1970) 'The provincial Stock Exchanges, 1830–1870', *Economic History Review*, 2nd series, 23.

King, W. T. C. (1936) *History of the London Discount Market*, London.

Kussmaul, A. (1981) *Servants in Husbandry in Early Modern England*, Cambridge.

Lambert, W. R. (1983) *Drink and Sobriety in Victorian Wales c.1820–c.1895*, Cardiff.

Langton, J. (1983) 'Liverpool and its hinterland in the late eighteenth century', in Anderson, B. L. and Stoney, P. M. S. (eds) *Commerce, Industry and Transport: Studies in Economic Change on Merseyside*, Liverpool.

Langton, J. (1984) 'The industrial revolution and the regional geography of England', *Transactions of the Institute of British Geographers*, new series, 9.

Langton, J. and Hoppe, G. (1983) *Town and Country in the Development of Early Modern Western Europe*, (IBG Historical Geography Research Group, publication 11), Norwich.

Lawton, R. (1968) 'Population changes in England and Wales in the later nineteenth century', *Transactions of the Institute of British Geographers*, 44: 55–74.

Lawton, R. (1977) 'Regional population trends in England and Wales, 1750–1971', in Hobcraft, J. and Rees, P. (eds) *Regional Demographic Development*, London.

Lawton, R. (1978a) 'Population and society 1730–1900', in Dodgshon, R. A. and Butlin, R. A. (eds) *An Historical Geography of England and Wales*, London.

Lawton, R. (1978b) *The Census and Social Structure*, London.

Lawton, R. (ed.) (1979) 'Mobility in nineteenth-century British cities', *Geographical Journal*, 145: 206–24.

Lawton, R. (1983) 'Urbanization and population change in nineteenth century England', in Patten, J. (ed.) *The Expanding City*, London.

Lee, C. H. (1979) *British Regional Employment Statistics 1841–1971*, Cambridge.

Lee, C. H. (1981) 'Regional growth and structural change in Victorian Britain', *Economic History Review*, 2nd Series, 33.

Leeson, R. A. (1979) *Travelling Brothers*, London.

Leith, D. (1983) *A Social History of English*, London.

Lenman, B. (1975) *From Esk to Tweed*, Glasgow.

Levitt, I. (ed.) (forthcoming) *Government and Social Conditions in Scotland, 1845–1945*.

Levitt, I. and Smout, C. (1979) *The State of the Scottish Working Class in 1843: A Statistical and Spatial Enquiry based on the Data from the Poor Law Commission Report of 1844*, Edinburgh.

Lewin, J. (1981) *British Rivers*, London.

Lewis, M. J. T. (1970) *Early Wooden Railways*, London.

Lindert, P. H. and Williamson, J. G. (1983) 'English workers' living standards during the Industrial Revolution: A new look', *Economic History Rweview*, 2nd series, 36.

Lockhart, D. G. (1980) 'The planned villages', in Parry, M. L. and Slater, T. R. (eds) *The Making of the Scottish Countryside*, London, 294–370.

Loudon, J. C. (1839) *An Encyclopaedia of Agriculture*, 4th edn, London.

Lunge, G. (1880) *A Theoretical and Practical Treatise on the Manufacture of Sulphuric Acid and Alkali*, London, 3 vols.

Macdonald, S. (1975) 'The progress of the early threshing machine', *Agricultural History Review*, 23: 63–77.

Mackinder, H. J. (1907) *Britain and the British Seas*, Oxford.

McLeod, Hugh (1973) 'Class, community and region: The religious geography of 19th century England', in Hill, M. (ed.)

Sociological Yearbook of Religion in Britain.

McLeod, Hugh (1974) *Class and Religion in the Late Victorian City*, London.

Manley, G. (1952) *Climate and the British Scene*, London.

Manley, G. (1974) 'Central England temperatures: monthly means 1659–1973', *Quarterly Journal of the Royal Meteorological Society*, 100: 389–405.

Mann, J. de L. (1971) *The Cloth Industry in the West of England from 1640 to 1880*, Oxford.

Mantoux, P. (1928) *The Industrial Revolution in the Eighteenth Century*, London.

Marsden, W. E. (1977a) 'Education and the social geography of nineteenth-century towns and cities', in Reeder, D. A. (ed.) *Urban Education in the Nineteenth Century*, London.

Marsden, W. E. (1977b) 'Historical geography and the history of education', *History of Education*, 6 (1).

Marsden, W. E. (1977c) 'Social environment, school attendance and educational achievement in a Merseyside town 1870–1900', in McCann, P. (ed.) *Popular Education and Socialization in the Nineteenth Century*, London.

Marsden, W. E. (1978) 'Variations in educational provision in Lancashire during the school board period', *Journal of Educational Administration and History*, 10 (2).

Marsden, W. E. (1979) 'Census enumerators' returns, schooling and social areas in the late Victorian town: a case study of Bootle', in Lowe, R. (ed.) *New Approaches to the Study of Popular Education 1851–1902*, Leicester, 16–33.

Marsden, W. E. (1982a) 'Diffusion and regional variation in elementary education in England and Wales 1800–1870', *History of Education*, 11 (3).

Marsden, W. E. (1982b) 'Schools for the urban power middle class: third grade or higher grade?' in Searby, P. (ed.) *Educating the Victorian Middle Class*, Leicester, 45–56.

Marsden, W. E. (1983) 'Ecology and nineteenth century urban education', *History of Education*, 23 (1).

Marshall, J. D. (1978) *The Old Poor Law, 1790–1930*, London.

Massey, D. (1984) *Spatial Divisions of Labour*, London.

Mather, F. C. (1972) *Chartism*, London.

Mathias, P. and Postan, M. M. (1978) *The Industrial Economies: Capital, Labour and Enterprise* (Cambridge Economic History of Europe, vol. 7, pt 1), Cambridge.

Mills, D. R. (1980) *Lord and Peasant in Nineteenth Century Britain*, London.

Minchinton, W. E. (1957) *The Trade of Bristol in the Eighteenth Century*, Bristol.

Minchinton, W. E. (1968) *Essays in Agrarian History*, 2 vols, Newton Abbot.

Mingay, G. E. (ed.) (1981) *The Victorian Countryside*, 2 vols, London.

Mitchell, B. R. and Deane, P. (1962) *Abstract of British Historical Statistics*, Cambridge.

Morgan, E. V. and Thomas, W. A. (1962) *The Stock Exchange: Its History and Function*, London.

Morgan, V. (1971) 'Agricultural wage rates in late eighteenth century Scotland', *Economic History Review*, 2nd series, 24.

Morris, R. J. (1971) 'The Friars and Paradise, an essay in the building history of Oxford 1801–1861', *Oxoniensia*, 36.

Mounfield, P. R. (1965) 'The shoe industry in Staffordshire', *North Staffordshire Journal of Field Studies*, 5: 74–80.

Mounfield, P. R. (1968) *The Footwear Industry of the East Midlands*, Nottingham.

Mounfield, P. R. (1970) 'An evaluation of census data for showing changes in industrial location', in Osborne, R. H., Barnes, F. A. and Doornkamp, J. C. (eds) *Geographical Essays in Honour of Professor K. C. Edwards*, Nottingham.

Mounfield, P. R. (ed.) (1972) 'The foundations of the modern industrial pattern', in Pye, N. (ed.) *Leicester and its Region* (British Association Handbook), Leicester.

Mounfield, P. R. (1977) 'The place of time in economic geography', *Geography*, 62: 268–85.

Mounfield, P. R. (1978a) 'The onset of the footwear industry in Northamptonshire and Leicestershire', in Palmer, M. (ed.) *The Onset of Industrialization*, Nottingham.

Mounfield, P. R. (1978b) 'Early technological innovation in the British footwear industry', *Industrial Archaeology Review*, 11 (2): 129–42.

Mulhall, M. G. (1903) *Dictionary of Statistics*, 45th edn, London.

Munn, C. W. (1981) *The Scottish Provincial Banking Companies 1747–1864*, Edinburgh.

Musson, A. E. (1976) 'Industrial motive power in the U.K., 1800–70', *Economic History Review*, 2nd series, 29.

Musson, A. E. (1980) 'The engineering industry', in Church, R. A. (ed.) *The Dynamics of Victorian Business: Problems and Perspectives to the 1870s*, London.

Musson, A. E. and Robinson, E. (1969) *Science and Technology in the Industrial Revolution*, Manchester.

Myrdal, G. (1957) *Economic Theory and Under-Developed Regions*, London.

Newham, H. E. C (1913) *Hull as a Coal Port*, Hull.

Newson, M. D. (1981) 'Mountain streams', in Lewis, J. (ed.) *British Rivers*, London, 59–89.

Nishimura, S. (1971) *The Decline of Inland Bills of Exchange in the London Money Market, 1855–1913*, Cambridge.

Orwin, C. S. and Whetham, E. H. (1971) *History of British Agriculture, 1846–1914*, Newton Abbot (orig. publ. London, 1964).

Overton, M. (1985) 'The diffusion of agricultural innovations in early modern England: turnips and clover in Norfolk and Suffolk, 1580–1740', *Transactions of the Institute of British Geographers*, n.s. 10: 205–21.

Overton, M. (1987) *Agricultural Revolution in England: the Transformation of the Rural Economy*, 1500–1830, Cambridge.

Parry, M. L. (1978) *Climatic Change, Agriculture and Settlement*, Folkestone.

Parry, M. L. and Slater, T. R. (eds) (1980) *The Making of the Scottish Countryside*, London.

Patmore, J. A. (1961) 'The railway network of Merseyside', *Transactions of the Institute of British Geographers*, 29: 233.

Paton, D. C. (1976) 'Drink and the temperance movement in nineteenth century Scotland', Ph.D. thesis, University of Edinburgh.

Pawson, E. (1977) *Transport and Economy: The Turnpike Roads of Eighteenth Century Britain*, London.

Pelling, H. (1957) *A Social Geography of British Elections, 1885–1910*, London.

Perkin, H. (1970) *The Age of the Railway*, London.

Perry, P. J. (ed.) (1973) *British Agriculture 1875–1914*, London.

Phillips, A. D. M. and Walton, J. R. (1975) 'The distribution of personal wealth in English towns in the mid-nineteenth century', *Transactions of the Institute of British Geographers*, 64.

Pollard, S. (1951) 'The decline of shipbuilding on the Thames', *Economic History Review*, 2nd series, 24.

Pollard, S. (1980) 'A new estimate of British coal production, 1750–1850', *Economic History Review*, 2nd series, 33, 212–35.

Pollard, S. and Robertson, P. (1979) *The British Shipbuilding Industry, 1870–1914*, Cambridge.

Pollins, H. (1971) *Britain's Railways: an Industrial History*, Newton Abbot.

Pooley, C. (1977) 'Migrants in mid Victorian Liverpool', *Transactions of the Institute of British Geographers*, n.s. 2.

Porritt, E. (1903) *Parliamentary Representation Before 1832*, 2 vols, Cambridge.

Porteous, J. (1977) *Canal Ports*, London.

Pressnell, L. S. (1956) *Country Banking in the Industrial Revolution*, Oxford.

Prothero, I. (1979) *Artisan Politics in Early Nineteenth Century London*, Baton Rouge, Louisiana.

Pryce, W. T. R. (1975) 'Patterns of migration and the evolution of cultural areas: cultural and linguistic frontiers in north east Wales, 1750–1851', *Transactions of the Institute of British Geographers*, 45.

Pryce, W. T. R. (1978a) 'Welsh and English in Wales: a spatial analysis based on the linguistic affiliations of parochial communities', *Bulletin of the Board of Celtic Studies*, 28.

Pryce, W. T. R. (1978b) 'Wales as a cultural region: patterns of change, 1750–1971', *Transactions of the Honourable Society of Cymmrodorion*.

Pryce, W. T. R. (1981) 'The Welsh language, 1751–1961', in Carter, H. and Griffiths, H. (eds), *The National Atlas of Wales*, Cardiff.

Pudney, J. (1975) *London's Docks*, London.

Purvis, M. (1986) 'Co-operative retailing in England, 1835–1850: Developments beyond Rochdale', *Northern History*.

Ravenstein, E. G. (1879) 'On the Celtic languages in the British Isles: a statistical survey', *Transactions of the Royal Statistical Society*, 42.

Read, D. (1964) *The English Provinces c.1760–1964*, London.

Reed, M. C. (ed.) (1969) *Railways in the Victorian Economy*, Newton Abbot.

Richardson, H. W. (1969) *Regional Economics: Location Theory, Urban Structure, and Regional Change*, ch. 13, London.

Richardson, T. and Watts, H. (1863) *Chemical Technology: Acids, Alkalis and Salt*, London.

Riddell, J. F. (1979) *Clyde Navigation: A History of the Deepening and Development of the River Clyde*, Edinburgh.

Riden, P. (1977) 'The output of the British iron industry before 1870', *Economic History Review*, 2nd series, 30 (3): 442–59.

Riden, P. (1980) 'The iron industry', in Church, R. A. (ed.) *The Dynamics of Victorian Business: Problems and Perspectives to the 1870s*, London, 63–86.

Roberts, B. K. (1979) 'Village plans in Britain', in Claude, J. (ed.) *Recherches de géographie rurale: Hommage au Professeur Frans Dussart*, Liège, 34–49.

Roberts, B. K. (1982) *Village Plans*, Princes Risborough.

Robson, B. T. (1969) *Urban Analysis: A Study of City Structure*, Cambridge.

Robson, B. T. (1973) *Urban Growth: An Approach*, London.

Rodgers, H. B. (1960) 'The Lancashire cotton industry in 1840', *Transactions of the Institute of British Geographers*, 28.

Rodgers, H. B. (1962) 'The changing geography of the Lancashire cotton industry', *Economic Geography*, 38.

Rolt, L. T. C. (1950) *The Inland Waterways of England*, London.

Rose, M. E. (1971) *The English Poor Law, 1790–1930*, Newton Abbot.

Rosenberg, N. (1963) 'Technological change in the machine tool industry 1840–1910', *Journal of Economic History*, 23.

Ross, H. M.(1904) *British Railways, their Organization and Management*, London.

Rubinstein, W. D. (1977) 'The Victorian middle classes: Wealth, occupation and geography', *Economic History Review*, 2nd series, 30.

Rubinstein, W. D. (1981) *Men of Property: The Very Wealthy in Britain since the Industrial Revolution*, London.

Saul, S.B. (1960) *Studies in British Overseas Trade, 1870–1914*, Liverpool.

Saul, S. B. (1967) 'The market and the development of the mechanical engineering industries in Britain, 1860–1914', *Economic History Review*, 2nd series, 20.

Saul, S. B. (1968) 'The machine tool industry in Britain to 1914', *Business History*, 10.

Saville, J. (1957) *Rural Population in England and Wales 1851–1951*, London.

Sayers, R. S. (1957) *Lloyds Bank in the History of English Banking*, Oxford.

Schofield, R. (1985) 'English marriage patterns revisited', *Journal of Family History*, 10: 2–20.

Shaw, G. and Wild, M. T. (1979) 'Retail patterns in the Victorian city', *Transactions of the Institute of British Geographers*, n.s. 4 (2): 278–91.

Shaw, J. (1984) *Water Power in Scotland 1550–1870*, Edinburgh.

Sheail, J. (1982) 'Underground water abstraction: indirect effects of urbanization on the countryside', *Journal of Historical Geography*, 8: 395–408.

Shields, J. (1947) *Clyde Built: A History of Shipbuilding on the Clyde*, Glasgow.

Shipley, S. (1971) *Club Life and Socialism in Mid-Victorian London*, History Workshop Pamphlet 5, Oxford.

Sigsworth, E. M. (1967) *The Brewing Trade during the Industrial Revolution: The Case of Yorkshire*, University of York, Borthwick Institute of Historical Research, Paper 31.

Skeat, W. W. and Nodal, J. H. (1877) *A Bibliographical List of Works that have been Published, or are Known to Exist in MS, Illustrative of the Various Dialects of English*, London.

Slaven, A. (1980) 'The shipbuilding industry', in Church, R. A., (ed.) *The Dynamics of Victorian Business: Problems and Perspectives to the 1870s*, London.

Smith, D. M. (1962) 'The cotton industry in the East Midlands', *Geography*, 47.

Smith D. M. (1963) 'The British hosiery industry at the middle of the nineteenth century: An historical study in economic geography', *Transactions of the Institute of British Geographers*, 32.

Snell, K. (1985) *Annals of the Labouring Poor*, Cambridge.

Southall, J. E. (1982) *Wales and her Language*, Newport, Mon.

Sparks, B. W. and West, R. G. (1972) *The Ice Age in Britain*, London.

Sparks, W. L. (1949) *Shoemaking in Norwich*, Norwich.

Stamp, J. C. (1916) *British Income and Property*, London.

Stamp, L. D. (1946) *Britain's Structure and Scenery*, London.

Stephens, W. B. (1973) *Regional Variations in Education during the Industrial Revolution 1780–1870*, University of Leeds Museum of Education.

Stephens, W. B. (1977) 'Illiteracy and schooling in the provincial towns, 1640–1870: a comparative approach', in Reeder, D. A. (ed.) *Urban Education in the Nineteenth Century*, London, 28–47.

Swann, D. (1960) 'The pace and progress of port investment in England, 1660–1830', *Yorkshire Bulletin of Economic and Social Research*, 12.

Swann, D. (1968) 'The engineers of English port improvement, 1660–1830', *Transport History*, 1.

Sykes, J. (1926) *The Amalgamation Movement in English Banking, 1825–1924*, London.

Tann, J. (ed.) (1981) *The Selected Papers of Boulton & Watt, Vol. 1: The Engines Partnership, 1775–1825*, London and Cambridge, Mass.

Taylor, A. M. (1964) *Gilletts: Bankers at Banbury and Oxford*, Oxford.

Temin, P. (1966) 'Steam and water power in the early 19th century', *Journal of Economic History*, 26.

Temple-Patterson, A. (1971) *A History of Southampton, 1700–1914*, 3 vols, London.

Thirsk, J. (1967) 'The Farming regions of England', in Thirsk, J. (ed.) *The Agrarian History of England and Wales, IV*, Cambridge.

Thirsk, J. (1984) *The Rural Economy of England*, London.

Thomas, S. E. (1935) *The Rise and Growth of Joint Stock Banking: Britain to 1860*, London.

Thomas, W. A. (1973) *The Provincial Stock Exchanges*, London.

Thompson, D. M. (1978) 'The Religious Census of 1851', in Lawton, R. (ed.) *The Census and Social Structure*, London.

Thompson, E. P. (1968) *The Making of the English Working Class*, Harmondsworth.

Thompson, F. M. L. (1968) 'The second agricultural revolution 1850–1880', *Economic History Review*, 2nd series, 21: 62–77.

Thorpe, H. (1964) 'Rural settlement', in Watson, J. W. and Sissons, J. B. (eds) *The Britsh Isles: A Systematic Geography*, London, 358.

Timperley, L. (1980) 'The pattern of landholding in eighteenth century Scotland', in Parry, M. L. and Slater, T. R. (eds) *The Making of the Scottish Countryside*, London, 137–54.

Tischler, S. (1981) *Footballers and*

Businessmen, New York.

Tomlinson, W. W. (1915) *The North Eastern Railway*, Newcastle.

Tribe, K. (1981) *Genealogies of Capitalism*, ch. 2, London.

Underdown, D. (1985) *Revel, Riot and Rebellion*, Oxford.

Vamplew, W. (1976) *The Turf*, London.

Vamplew, W. (1982) 'The economics of a sports industry: Scottish gatemoney football, 1890–1914', *Economic History Review*, 2nd series, 35.

Vicinus, M. (1974) *The Industrial Muse: A Study of Nineteenth Century British Working Class Literature*, London.

von Tunzelmann, G. N. (1978) *Steam Power and British Industrialization to 1860*, Oxford.

Wadsworth, A. P. and Mann, J. de L. (1931) *The Cotton Industry and Industrial Lancashire 1600–1780*, Manchester.

Wakelin, M. F. (1977) *English Dialects: An Introduction*, London.

Walters, R. C. S. (1936) *The Nations's Water Supply*, London.

Walton, J. R. (1978) 'Mechanisation in agriculture: a study of the adoption process', in Fox, H. S. A. and Butlin, R. A. (eds) *Change in the Countryside*, London, 23–42.

Walton, J. R. (1984) 'The diffusion of the improved shorthorn breed of cattle in Britain during the eighteenth and nineteenth centuries', *Transactions of the Institute of British Geographers*, n.s. 9: 22–36.

Ward, D. (1980) 'Environs and neighbourhoods in two nations', *Journal of Historical Geography*, 6.

Ward, R. C. (1981) 'River systems and river régimes', in Lewis, J. (ed.) *British Rivers*, London, 1–34.

Ward, W. R. (1972) *Religion and Society in England, 1790–1850*, London.

Warren, K. (1980) *Chemical Foundations: The Alkali Industry in Britain to 1926*, Oxford.

Watson, J. W. and Sissons, J. B. (1964) *The British Isles: A Systematic Geography*, London.

Webb, B. and S. (1920) *The History of Trade Unionism, 1666–1920*, 2nd edn, London.

Weber, A. F. (1899) *The Growth of Cities in the Nineteenth Century*, New York.

Weir, R. B. (1980) 'The drink trades', in Church, R. A. (ed.) *The Dynamics of Victorian Business*, London.

Wigley, T. L. M., Lough, J. M. and Jones, P. D. (1984) 'Spatial patterns of precipitation in England and Wales and a revised, homogenized England and Wales precipitation series', *Journal of Climatology*, 4 (1), 25.

Wilde, P. D. (1975) 'Locational problems in the English silk industry in the mid-nineteenth century', in Phillips, A. D. M. and Turton, B. J. (eds) *Environment, Man and Economic Change: Essays presented to S. H. Beaver*, London.

Willan, T. S. (1936) *River Navigation in England, 1600–1750*, Manchester.

Wilson, G. B. (1940) *Alcohol and the Nation*, London.

Withers, C. J. W. (1984) *Gaelic in Scotland 1698–1981: The Geographical History of a Language*, Edinburgh.

Withers, C. J. W. (1985) 'Kirk, club and social change: Gaelic chapels, Highland Societies and the urban Gaelic subculture in eighteenth-century Scotland', *Social History*, 10.

Women and Geography Study Group of the IBG (1984) *Geography and Gender*, London.

Woods, R. I. (1982) *Theoretical Population Geography*, London, 102–30.

Wren, W. J. (1976) *Ports of the Eastern Counties*, Lavenham.

Wright, J. (1896) *Bibliography of English Dialects*, Bodleian Library, 2585.

Wrigley, E. A. (1967) 'A simple model of London's importance in changing English society and economy, 1650–1750', *Past and Present*, 37.

Wrigley, E. A. and Schofield, R. (1981) *The Population History of England 1541–1871*, London.

Yeats, J. (1871) *The Natural History of the Raw Materials of Commerce*, London.

Yeats, J. (1890) *Map Studies of the Mercantile World*, London.

Yelling, J. (1977) *Common Field and Enclosure in England 1450–1850*, London.